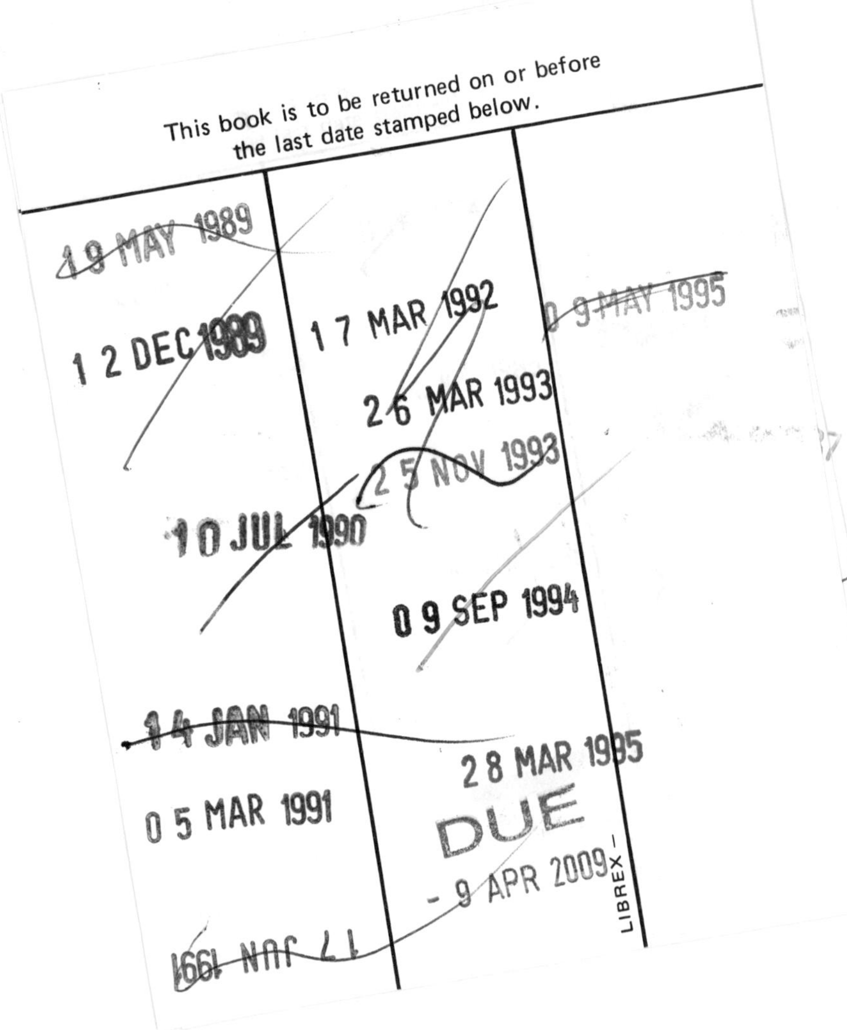
This book is to be returned on or before
the last date stamped below.
19 MAY 1989
12 DEC 1989
10 JUL 1990
14 JAN 1991
05 MAR 1991
17 JUN 1991
17 MAR 1992
26 MAR 1993
25 NOV 1993
09 SEP 1994
28 MAR 1995
DUE
- 9 APR 2009
09 MAY 1995
LIBREX

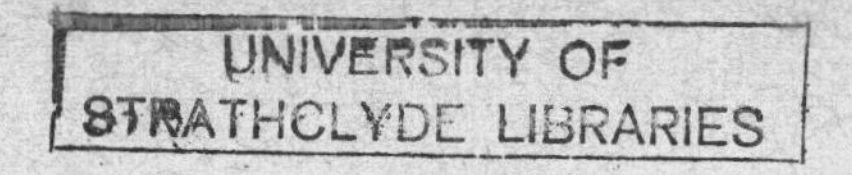

The Open University

A Second Level Course

INSTRUMENTATION

Units 11, 12 and 13

Noise in Instrumentation Systems

Prepared by the Course Team

THE OPEN UNIVERSITY PRESS

The Instrumentation Course Team

D. I. Crecraft	(Chairman)
G. R. Alexander	(Electronics)
G. Bellis	(Scientific Officer)
G. E. Harland	(Staff Tutor)
S. L. Hurst	(Visiting Lecturer)
D. S. Jackson	(BBC)
A. B. Jolly	(BBC)
R. D. R. Kyd	(Editor)
R. Loxton	(Electronics)
G. S. Martin	(Electronics)
J. Monk	(Electronics)
G. Smol	(Electronics)
J. J. Sparkes	(Electronics)
C. M. Walker	(Course Assistant)

The Open University Press
Walton Hall, Milton Keynes

First published 1974

Designed by the Media Development Group of the Open University.

Printed in Great Britain by
Staples Printing Group
at St Albans.

ISBN 0 335 02954 X

This text forms part of an Open University course. The complete list of units in the course appears at the end of this text.

For general availability of supporting material referred to in this text, please write to the Director of Marketing, The Open University, PO Box 81, Milton Keynes, MK7 6AT.

Further information on Open University courses may be obtained from the Admissions Office, The Open University, PO Box 48, Milton Keynes, MK7 6AB.

1.1

Aims

The aims of Units 11, 12 and 13 are:

Unit 11

1 To explain the problems and considerations which arise when signals are interfered with by extraneous voltages.

2 To introduce three case studies, one of which will be examined in this unit.

3 To explain how to quantify a finite segment of a signal.

4 To explain the terms 'random waveform', 'mean value', 'r.m.s. value', 'standard deviation' and 'probability density function'.

5 To introduce the sampling theorem and to show some of its implications.

6 To explain the term 'Gaussian noise' and to show, with the aid of a simple example, why many random distributions are in fact Gaussian.

7 To show the effect of signal averaging upon noise.

8 To demonstrate how signal averaging can be used.

Unit 12

9 To explain the concept of a continuous spectrum.

10 To describe some ways in which the power density spectrum of a signal can be measured.

11 To describe some instrumentation systems in which the system's bandwidth has been 'matched' to the signal's bandwidth.

12 To explain the autocorrelation function and some of its uses.

13 To explain the idea that modulation is helpful in combating a noisy channel.

14 To outline the reasons why frequency modulation and pulse code modulation can be made more resistant to a noisy channel than amplitude modulation.

Unit 13

15 To explain the use of error-detecting and error-correcting codes.

16 To describe a case study in which an error-correcting code is used.

17 To explain the cross-correlation function and to describe some of its uses.

18 To describe a case study in which cross-correlation techniques are used to measure the velocity flow rate of a dirty fluid.

Objectives

Content areas	Objectives
	When you have completed these units you should be able:
Unit 11, Section 2	
1 Terminology: Signal, wanted signal, noise, waveform, filter, low-pass filter, high-pass filter, band-pass filter, bandwidth	To distinguish between true and false statements concerning each of these terms.
Unit 11, Section 3	
2 Terminology: Thermal noise, shot noise, stellar noise, burst noise	To distinguish between true and false statements concerning each of these terms.
Unit 11, Section 5	
3 Terminology: Sampling, quantifying a waveform, random waveform, deterministic waveform	To distinguish between true and false statements concerning each of these terms.
4 Use of Fourier components to specify a waveform	To explain how Fourier components can be used in this way.
5 Demonstration of how a band-limited waveform, whose spectrum extends from zero frequency to f_c and whose duration is T_o, can be specified by $2f_cT_o$ quantities	To explain and demonstrate this fact.
6 Reconstitution of a waveform from the appropriate number of samples of it	To describe how this can be done.
Unit 11, Section 6	
7 Mean values, r.m.s. value and standard deviation	To calculate these values of a waveform from a set of samples of it.
8 Histogram	To draw this from a set of sampled data of a waveform.
9 Histogram and probability density function	To explain the difference between these terms.
Unit 11, Section 7	
10 Terminology: Binomial distribution, Gaussian distribution, signal averaging	To distinguish between true and false statements concerning each of these terms.
11 The Gaussian probability density function	To describe the properties of this function and explain why it describes many probability distributions of naturally occurring quantities.
12 Signal averaging	To explain how this process improves signal-to-noise ratio and state by what factor the ratio is improved when the noise is Gaussian.
Unit 11, Section 8	
13 Terminology: Averaging or synchronous detection	To distinguish between true and false statements concerning these terms.
Unit 12, Section 2	
14 Terminology: Aperiodic, Fourier components, spectrum, wave analyser, variable-frequency narrow-band filter, bandwidth, attenuation, line spectrum, con-	To distinguish between true and false statements concerning each of these terms.

	tinuous spectrum, power density spectrum, wide-band filter, white noise, wide-band noise, narrow-band noise, $1/f$ noise	
15	Spectral representation of an aperiodic waveform	To appreciate why the power density spectrum of an aperiodic waveform is often treated as being continuous.
16	Use of a spectrum analyser to measure the power density spectrum of a signal	To explain the need for a spectrum analyser's readings to be quoted in terms of power density.
17	The power of a signal	To understand what the area under the power density spectrum represents.
18	Different types of spectrum analysers	To appreciate some of the other methods for measuring a signal's spectrum.
19	Problems of measuring the spectrum of a signal	To outline some of the problems of this measurement.

Unit 12, Section 3

20	Terminology: Band pass, band stop, cut-off frequency, shielding, notch filter, decibel (dB)	To distinguish between true and false statements concerning each of these terms.
21	The decibel unit	To calculate the ratio of two voltages and express the result in decibels.
22	Examples of instrumentation systems whose bandwidth is matched to that of the signal	To appreciate the need to match an instrumentation system's bandwidth to that of the signal.

Unit 12, Section 4

23	Terminology: Autocorrelation function, autocorrelation coefficient, autocorrelator, parametric time, chain code, pseudo-random binary sequence	To distinguish between true and false statements concerning each of these terms.
24	Autocorrelation	To obtain the autocorrelation function of a waveform from a set of sample values of the waveform.
25	Autocorrelation function for different types of waveforms	To recognize the difference between the autocorrelation functions of a random waveform and an aperiodic waveform.
26	Autocorrelation function and power density spectrum	To explain the relationship between the autocorrelation function of a signal and the power density spectrum of a signal.
27	Shape of autocorrelation functions	To appreciate the information about the spectrum of a signal that can be deduced from an examination of the shape of its autocorrelation function.
28	Autocorrelation as a means of detecting signals which appear to be hidden by noise	To appreciate the usefulness and the limitations of using autocorrelation as a means of detecting signals which are hidden by noise.

Unit 12, Section 5

29	Terminology: Noisy channel, modulation, carrier, demodulation, amplitude modulation (AM), frequency modulation (FM), pulse code modulation (PCM), sidebands, instantaneous frequency, quantization error, quantization noise, quantum level	To distinguish between true and false statements concerning each of these terms.
30	Amplitude modulation, frequency modulation and pulse code modulation	To explain the meaning of and the difference between amplitude modulation, frequency modulation and pulse code modulation.
31	The bandwidth needed for AM, FM and PCM	To describe the factors involved in estimating the required bandwidth of AM, FM and PCM.

32 Demodulation of FM	To explain why the signal-to-noise ratio at the output of an FM detector increases when the deviation of the FM waveform is increased.
33 PCM error rate	To explain qualitatively the effect of noise upon PCM.
Unit 13, Section 2	
34 Terminology: Coding, redundancy, vocabulary, error, message, message word, code word, redundant code word, even and odd parity, parity check code, ARQ code, error-detecting code	To distinguish between true and false statements concerning each of these terms.
35 Error-detecting codes	To explain the action of two error-detecting codes.
Unit 13, Section 3	
36 Terminology: Error-correcting codes, chain codes, modulo-2 addition, shift register, EXCLUSIVE-OR gate, pseudo-random sequence	To distinguish between true and false statements concerning each of these terms.
37 Error-correcting codes	To explain the principle of error-correcting codes.
38 Chain codes	To produce a chain code of a required sequence length using the rules for producing chain codes.
39 Properties of chain codes	To appreciate the cyclic properties of a chain code and to predict the length of a chain-code sequence given the number of message digits.
Unit 13, Section 4	
40 Terminology: Cross-correlation coefficient	To distinguish between true and false statements concerning this term.
41 Decoding a chain code	To calculate cross-correlation coefficients and to use them to decode a chain code which contains errors.
42 Error-correcting capability of a chain code	To estimate how many errors a chain code of a given length can correct and how many it can detect.
Unit 13, Section 5	
43 The Mariner 9 space probe	To explain the need for the use of error-correcting codes in the Mariner space probe.
44 Mariner 9's error-correcting code	To explain some of the differences between this code and an ordinary chain code.
45 The ranging of Mariner 9	To explain how pseudo-random sequences were used in ranging the probe.
Unit 13, Section 6	
46 Cross-correlation function	To appreciate some of the uses of the cross-correlation function.
Unit 13, Section 7	
47 Correlation flow meter	To describe how flow rates can be measured using cross-correlation techniques and appreciate some of the problems involved.
48 Cross-correlator	To explain the operation of a cross-correlator specially designed for measurements of flow rate.
49 Impulse response	To explain what is meant by the impulse response of a system.
50 Measurement of impulse response	To explain how the impulse response of a system can be measured using cross-correlation techniques.

Study guide for Unit 11

The following list is a guide to how you might divide your study time between the various components associated with this unit:

Studying the text	7 hours
Answering the self-assessment questions	$1\frac{1}{4}$ hours
Watching the television programme 'Signals and noise'	25 minutes
Listening to the radio programme 'Signal statistics'	20 minutes
Reading paper 17 in the Course Reader*	$\frac{1}{4}$ hour
Preparatory work for the assignments associated with Units 11, 12 and 13	$\frac{3}{4}$ hour

You need only read quickly through the Course Reader paper, as you are advised to read it again more thoroughly when you have completed Unit 13.

Study guide for Unit 12

The following list is a guide to how you might divide your study time between the various components associated with this unit:

Studying the text	7 hours
Answering the self-assessment questions	1 hour
Preparatory work for the assignments associated with Units 11, 12 and 13	1 hour

Study guide for Unit 13

The following list is a guide to how you might divide your study time between the various components associated with this unit:

Studying the text	6 hours
Answering the self-assessment questions	$1\frac{1}{4}$ hours
Watching the television programme 'Modulation and noise'	25 minutes
Listening to the radio programme 'Signals in noise'	20 minutes
Re-reading paper 17 in the Course Reader	20 minutes
Completing the assignments associated with Units 11, 12 and 13	1 hour 40 minutes

**Hunter, J. J. and Crecraft, D. I. (1973)* Instrumentation, *Open University Press/Holmes McDougall.*

Unit 11

Contents

Structure of Unit 11

This diagram summarizes the unit as a whole. The page numbers show where the summaries of each section can be found

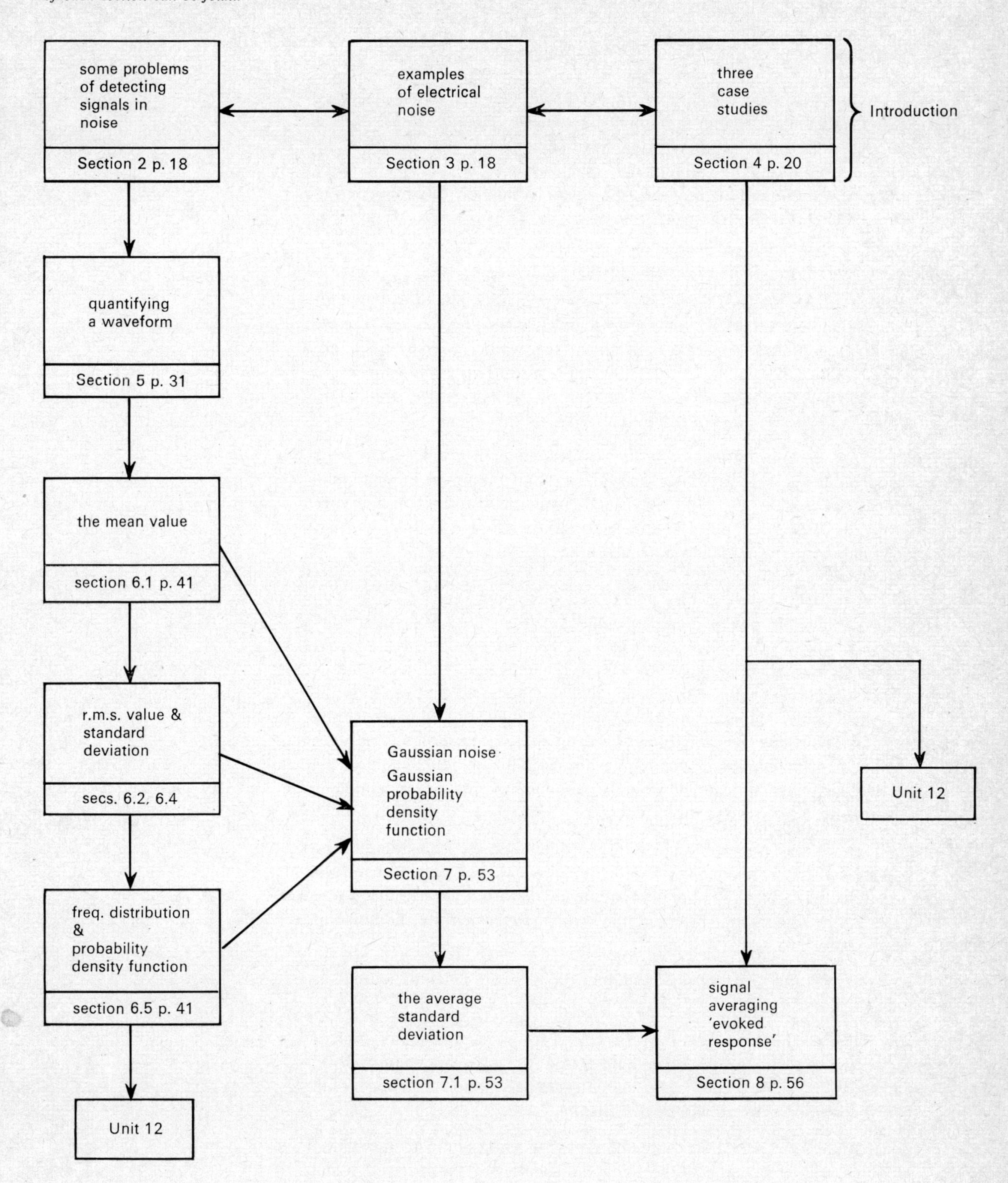

Introduction

So far in this course we have always assumed that the signal emerging from a transducer or from the amplifiers etc. connected to it was caused entirely by the measurand. We now have to consider what can be done when at any instant the value of the signal may be the sum of the effect of the measurand and of the value of a signal from some other interfering source. Evidently, in order to determine the value of the measurand from measurements of the signal voltage it is necessary to remove, or at least to know the value of, the interfering voltage.

There are many kinds of noise just as there are many kinds of instrumentation signal, and although there are occasions when there is nothing we can do about separating the noise from the wanted signal, there are several techniques which can be used in appropriate circumstances to improve the *signal-to-noise ratio*. The task of the next three units is to explain and justify some of these techniques.

There is a certain amount of 'noise' associated with every signal – the background hiss of your radio, the buzzing and humming in your telephone, the speckles on your television screen are examples of the consequences of electrical noise – so the assumption adopted up till now, that we could ignore it, needs some justification.

How do you think it can be justified?

The justification is simply that we have always assumed that the noise is, or can be made, negligible compared with the signal. Indeed, overcoming noise by increasing the amplitude of the signal is a common strategy not only in instrumentation systems. We all shout to be heard in cocktail parties, factories, canteens or wherever the background noise cannot be ignored, thus overcoming the interference.

But it is not always possible or desirable to increase the magnitude of the wanted signal to a value at which any interference is negligible. It is to such situations that Units 11, 12 and 13 are devoted, with a view to discussing what can be done about receiving the wanted signal despite the noise or interference.

Units 11, 12 and 13 form a continuous block. Unit 11 is structured as follows:

First I discuss the meanings of certain key words we shall be using, such as 'signal', 'waveform' and 'noise', illustrated by practical examples. In particular, some of the kinds of interference which can occur in instrumentation systems will be described.

Then I shall introduce three quite different case studies in instrumentation in which interference affects the signal sufficiently for the received signal to be inadequate for the purpose for which it was intended.

As the subject is developed the various methods of dealing with interference signals will be described with reference to these three case studies.

A good deal of the material is concerned with the properties of random signals and how they can be specified. An understanding of their properties is essential to an understanding of how to deal with them.

Section 2

Signals in instrumentation

2.1 Signals

Throughout the preceding units the term 'signal' has been frequently used to stand for the time-varying voltage (or current) used for carrying information about the value of a measurand. I am going to continue to use it in this sense, but it must be understood that from now on there can be some ambiguity about the precise meaning of the word. This ambiguity is normal and inevitable since it is now a part of our language but it must be appreciated and accepted if it is not to cause you trouble.

1 On the one hand we use the word 'signal' for the output of a transducer or for the input of an amplifier, for example, whether or not noise is part of this signal, as in Figure 1.

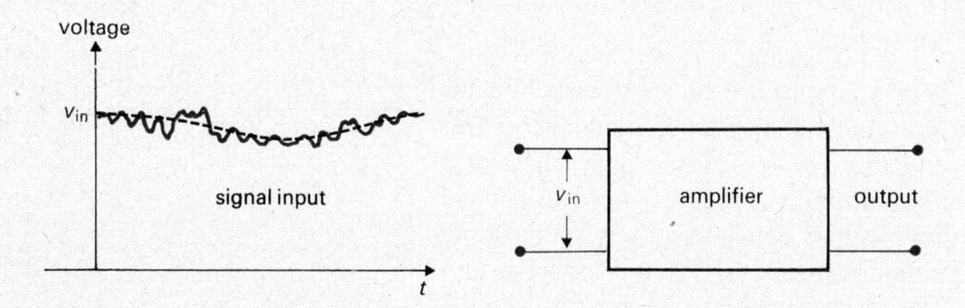

Figure 1 An input signal to an amplifier. The 'signal' is sometimes separated into two parts: the 'wanted signal' and the noise

2 On the other hand we frequently use the phrase 'signal-to-noise ratio', which refers to *two* components of the input to an amplifier. One part is the 'wanted', or 'required', voltage variation, the other being unwanted interference, or noise. In Figure 1 you can see that the slowly varying waveform might be the wanted signal and the random ripple the noise – or vice versa.

Thus the 'signal' is sometimes used to describe the whole signal and sometimes used to describe that part of the signal which is carrying the required information.

In many situations and discussions the two meanings are in any case practically synonymous since the noise is negligible. In Units 11, 12 and 13, at least, they are not synonymous because we are concerned with how to deal with a signal containing noise. Nevertheless, it is possible to use the words unambiguously even when both meanings are used side by side.

However, when it is necessary to resolve an ambiguity of expression I shall use the phrase 'wanted signal' to refer to the message-carrying part of an overall signal.

Briefly, then,

Signal=wanted signal+noise.

But where there is no danger of ambiguity I shall use the word 'signal' to stand for the 'wanted signal', so that under these circumstances 'signal' refers to that which carries the information.

All this just makes explicit the common usage of the word.

2.2 Waveforms

The word 'waveform' in many contexts is synonymous with the word 'signal', but not always. Although we use the word 'signal' to describe a time-varying voltage which contains *some* noise, it is not usual to use the word when there is no information. Thus, we speak of a sinusoidal *waveform*, but to refer to a sinusoidal *signal* is not helpful, since a pure sinusoid cannot carry information. So a signal can be thought of as a waveform which carries information. Again, random noise is a time-varying voltage which is not a signal because it too carries no information, but any finite segment of it has some particular waveform. Indeed, it is often called a random waveform. So this is one difference in usage of 'signal' and 'waveform'.

waveform **signal**

Now it is always true in instrumentation, if only because instruments do not respond instantaneously, that we observe finite durations of a signal, not instantaneous values. The instantaneous value of a signal is its value at a particular instant – a point on the graph of voltage against time. This value is in part due to the measurand and in part due to any noise present. On their own, that is, considering only this instant, these two parts are not separable. The techniques for separating them depend on taking averages of one kind or another, as we shall see, and this of course involves a non-zero duration of the signal.

Obviously, too, averaging times cannot go on for ever, so an instrumentation system always deals with a finite *observation time*. I shall refer frequently to this time, and shall reserve the symbol T_o for it. The subscript o stands for observation. The observation time may be microseconds or less or may extend over seconds or hours. It may depend on the response time of a meter or upon the amount of data recorded by a pen recorder or fed into a computer, but there must always be some time period involved. In general I shall refer to the signal within this observation time as 'a waveform', rather than as a signal. Thus a waveform in this sense can be thought of as part of a signal. But again common usage of the words makes no clear distinction of this kind.

observation time

Suppose you observe the output of a signal generator on a pen recorder or oscilloscope and you see the waveform ot Figure 2(a) – a sinusoidal waveform. You would no doubt infer that the *signal* generated by the instrument was also sinusoidal. But I want you to realize that this is no more than an inference made by *you* .You cannot tell from looking at the waveform whether or not the generator was switched on just before the observation time and switched off again immediately afterwards, whether its output was automatically varied, whether the waveform was part of a slowly decaying sinusoid, and so on.

The waveform of limited duration is all you actually observe or know about.

It is necessary to maintain a clear distinction in your mind between what you *observe* and what you *infer* from your observation. The waveform is what you observe. The properties of the rest of the signal are inferred on the basis of certain assumptions. For example, you have to assume something about the constancy of the generator if you infer that its output *signal* is sinusoidal, when all you have observed is a short sinusoidal waveform.

A normal oscilloscope display of a sinusoidal signal is often, of course, a repeated display of this particular waveform. The more times it is repeated the more convinced you become that the *signal* is sinusoidal. But sinusoidal signals are a very special case. No instrumentation signal carrying information is continuously sinusoidal. Only departures from a pure sinusoid carry information; either observations of when it is switched on and off or of different wave shapes are of interest. Steady unvarying

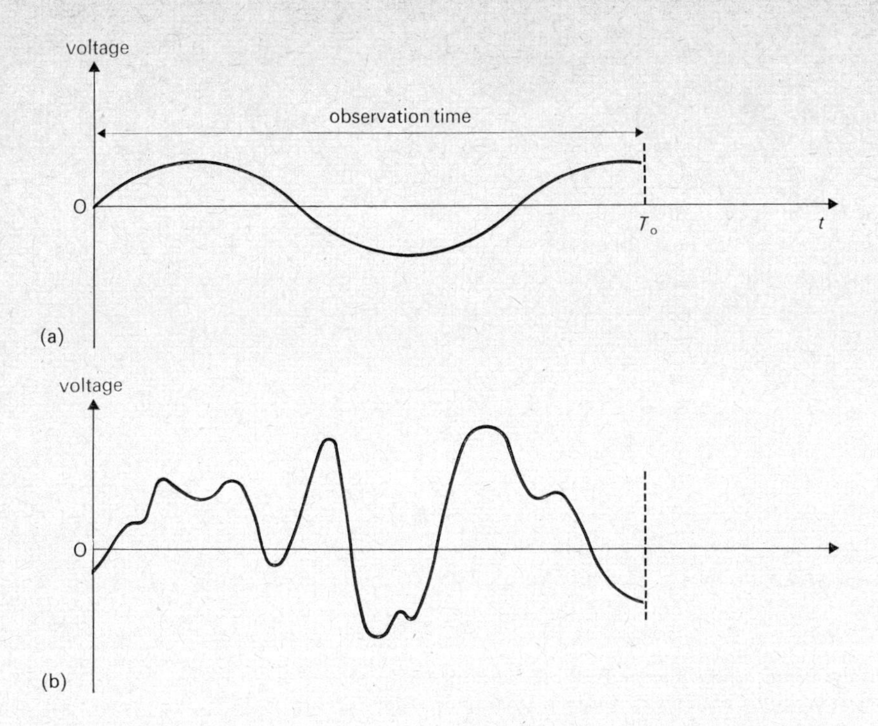

Figure 2 Two examples of waveforms: (a) a sinusoid; (b) a random waveform

sinusoids are only of value as test signals to ensure that the system is working properly. So the idea of a repetitive sinusoidal display is not of much relevance in instrumentation. It is waveforms, or segments of a signal, that we observe. The properties of signals as a whole can only be inferred.

2.3 Bandwidth

bandwidth

You have seen in earlier units how it is possible to describe signals in terms of their spectra, that is, in terms of their frequency components. I shall be using the term *bandwidth* to refer to the range of frequencies occupied by the frequency components of a signal or waveform. I shall also use it to refer to the range of frequencies that an instrumentation system can handle. Thus the bandwidth of a hi-fi amplifier may be about 15 kHz, meaning that it can handle components with frequencies between a few tens of hertz and 15 kHz. Similarly, the bandwidth of recognizable speech is about 3 kHz, extending from, say, 300 Hz to 3.3 kHz. Again, the bandwidth of a television signal is 5.5 MHz, extending from, say, 650 MHz to 655.5 MHz.

In all instrumentation systems where noise may be significant it is normal to ensure that the frequency characteristics of the channel, the amplifier, the indicator, etc. are the same as those of the signal.

We have seen in the previous units how it is necessary, in order to avoid distortion of the signal, to ensure that the range of frequencies covered by the system is sufficient. That is, the system bandwidth is no less than that of the signal.

I must now emphasize that in order to ensure that interference is not unnecessarily large the bandwidth of the system should be *no more* than that of the signal. If it is, the system may accept more noise than it need.

Thus the upper and lower cut-off frequencies of the system should be the same as those of the signal. The bandwidth and the actual frequency range must be the same for each.

An obvious example of this is the tuning of a radio. Thousands of signals reach the aerial of your radio, yet you usually manage to listen to just one programme at a time. All the unwanted signals – the medium-wave, long-wave, short-wave, television, etc. transmissions – are interference to the one signal you are wanting to detect. So included as part of the receiver is an electrical circuit which has an upper frequency limit which is about 8 kHz higher than the lower frequency limit. Thus only about 8 kHz of bandwidth can enter the receiver. All the rest is removed. When you tune the radio the centre frequency of the band passed by this circuit is adjusted so that it just matches that of the signal. Then all the unwanted signals are reduced, normally to a negligible level. See Figure 3.

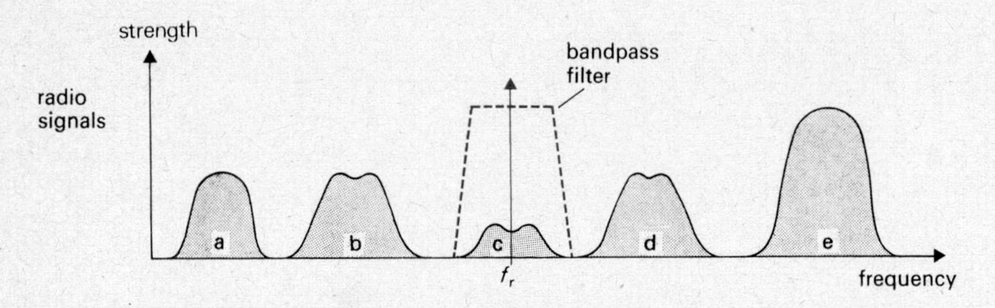

Figure 3 The various radio transmissions impinging upon an antenna are indicated and labelled (a–e). Tuning a radio consists of adjusting the centre frequency of a band-pass filter. As shown, the filter accepts only signal (c), although it is the weakest, and rejects all the others

What are circuits of this kind called?

You may recall from Unit 1 that they are called *filters*.

A filter in electronics is a circuit designed to change the spectrum of a signal. A circuit which attenuates the high frequencies more than the low frequencies is called a *low-pass filter*. A circuit which does the opposite is called a *high-pass filter*. A *band-pass filter* attenuates the high and low frequencies more than the frequencies lying within the pass band. The bandwidth of a filter is normally the range of frequencies which are not attenuated by more than 3 dB – that is, by a factor of no more than 0.7.

low-pass filter

high-pass filter **band-pass filter**

Some electronic millivoltmeters are provided with a switch with which to select the required bandwidth of the instrument. Here the low-frequency end of each bandwidth is zero frequency, and the bandwidth selector switch simply selects the upper cut-off frequency of a low-pass filter. The purpose of this is to allow you to filter out any noise which lies outside the range of frequencies occupied by the wanted signal.

Thus in the design of instrumentation systems in which noise or interference may be a problem it is necessary to discover, or to decide, what bandwidth the wanted signal will occupy and to design your system so that its bandwidth is as close to that of the signal as possible.

The best approach to discovering or deciding on the appropriate bandwidth depends upon the task involved. A knowledge of the characteristics of the transducer may tell you the highest-frequency component present. A knowledge of the general behaviour of the measurand may tell you the frequency limits you need to consider. You may not be concerned to follow the very rapid or the very slow variations of the signal because they do not affect the problem in hand, in which case you can *decide* to limit the system bandwidth, to filter out some of the signal and thus also to reduce the interference due to noise.

The point is that however the bandwidth of the wanted signal is obtained' the system bandwidth should be designed to match it – neither too much, nor too little.

In most of the examples we shall be discussing the lower cut-off frequency of the signal is taken to be zero. In Unit 12, however, we shall be considering how signals can be altered to reduce the effect of noise upon them, and some of these will have both a lower cut-off frequency and an upper one.

SAQ 1

SAQ 1

Figures 4(a–c) show the frequency response of three kinds of filter. Match each filter to the correct one of the following descriptions:

A low-pass filter,
B high-pass filter,
C band-pass filter,
D band-stop filter.

Figure 4(d) shows the spectrum of a signal. The spectrum extends from zero to a cut-off frequency f_c. Sketch the spectrum of this signal after it has been passed through a low-pass filter which has a sharp cut-off at $\frac{1}{2}f_c$.

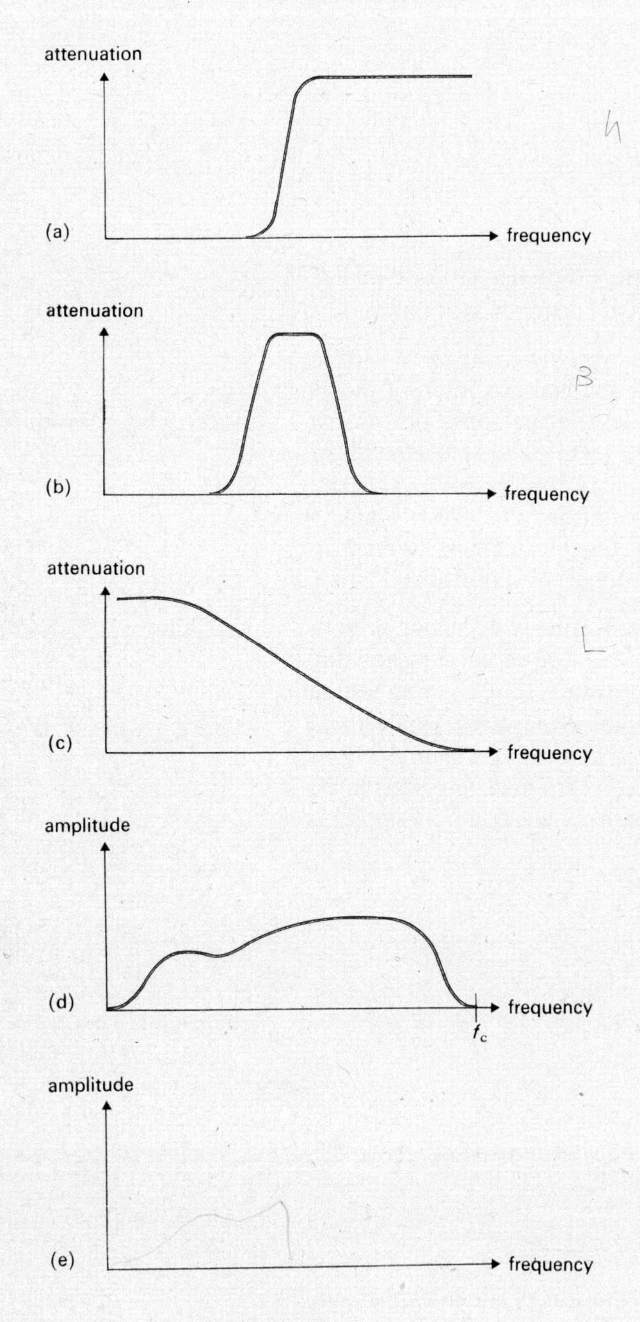

Figure 4 Diagram for SAQ 1

2.4 Summary

We shall be returning to a discussion of a number of these points in the following pages. This introduction has attempted to establish that:

1 There is an ambiguity in the meaning of the word 'signal'. That is,

Signal = wanted signal + noise

but 'wanted signal' is often called 'signal' if the ambiguity is not significant.

2 An instrumentation system deals with a finite duration of a signal. This duration is called the *observation time*. We shall usually refer to the signal within the observation time as a waveform.

3 The pass band of an instrumentation system should, as a first step towards reducing the effect of noise, be made the same as that of the signal.

Section 3

Examples of electrical noise

Before proceeding with a consideration of the properties of noise and of random signals I shall first briefly describe some common types of noise.

A type of noise which, to some extent, affects every electrical signal used in measurement is the random noise due to the thermal excitation of atoms and electrons in electrical circuits. It is called *thermal noise* or *Johnson noise*. At temperatures other than absolute zero these movements of electrons produce voltage fluctuations between any two points in a circuit. These fluctuations within the relevant bandwidth may be of the order of millivolts or microvolts, depending upon the circuit resistance, and may be quite comparable to the output of certain types of transducer.

thermal noise, Johnson noise

Shot noise is the name given to a type of electrical noise occurring in both vacuum tubes and transistors and so is another type of noise originating from components contained in electronic apparatus. In vacuum tubes the approach and arrival of each electron at the anode gives an electrical impulse and, since there are fluctuations in the rate at which electrons arrive, the instantaneous value of the electric current which flows fluctuates, even though the average value, as measured by a meter, for example, is constant. This is illustrated in Figure 5.

shot noise

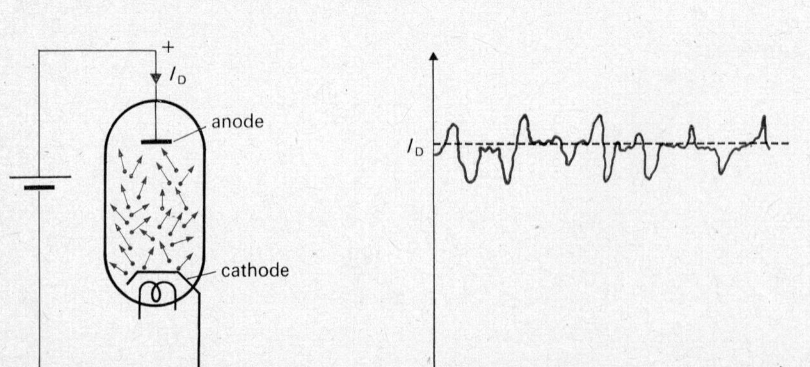

Figure 5 A thermionic diode and the variation with time of its current I_D due to the random flow of electrons from cathode to anode

In transistors the shot noise occurs because of the random nature of the diffusion of the discrete charge carriers within the semiconductor material used in the construction of transistors.

There are many other kinds of noise which affect electrical apparatus. The transients due to the switching of electric currents in a system, for example, can give rise to fluctuating voltages in other parts of the system. Some of these transients can be of relatively long duration and in such cases are referred to as *burst noise*. Burst noise due to other causes affects radio transmission paths and telephone lines and can submerge quite large portions of the transmitted signal. Lightning is a source of burst noise which can obliterate a second or more of a signal transmitted either by radio or by coaxial cable.

burst noise

Stellar noise, the radiation caused by processes of energy conversion in stars, is a subject of study for radio astronomers but it is also a source of interference affecting the transmission of information from satellites and space probes (see Figure 7).

stellar noise

These are some of the kinds of random noise which can be encountered. But not all interference is random – although the word 'noise' is normally reserved for interference which is random, rather than periodic, in nature.

A very common form of interference is 50 Hz 'hum' due to 'pick-up' from the a.c. mains. This too can interfere with measurements and may have to be removed before the measurand can be detected. Much of it can sometimes be balanced out using a differential input amplifier, as indicated in Figure 6. Filtering using a narrow-band band-stop filter can further reduce it if necessary.

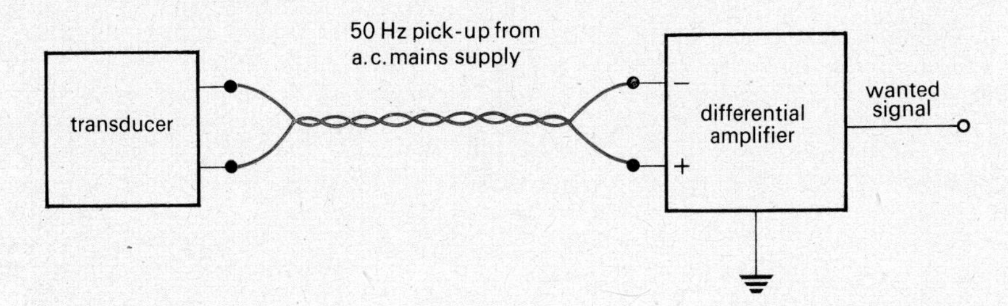

Figure 6 Pick-up from the a.c. mains and its reduction by a differential amplifier. The a.c. mains is normally 240 V, *much larger than the transducer output, which may be only a few microvolts or less. Thus it is possible for the output wires to 'pick up' a voltage which is not negligible compared with the wanted signal. However, if two intertwined wires connect the transducer to a differential amplifier, each wire will pick up almost the same* 50 Hz *voltage, so that it will be largely balanced out at the differential input. The amplifier will then amplify the wanted signal primarily. This is dealt with further in Unit 16*

Three case studies

The following examples of instrumentation problems are concerned with the detection and measurement of signals in the presence of a significant amount of noise.

4.1 The Mariner space probe

The Mariner space probe was launched with the specific purpose of taking photographs of Mars and for measuring certain properties of the planet. The task of converting the delicate measurements involved into electrical signals is a problem of transducer design which is *not* the question that concerns us here. The point of interest is the problem of receiving, without error, the signals transmitted back to Earth. The signals produced by the probe suffer an enormous degree of attenuation on their way to Earth, so much so that on reaching us they are as weak as the stellar noise which reaches us from outer space. (See Figure 7.)

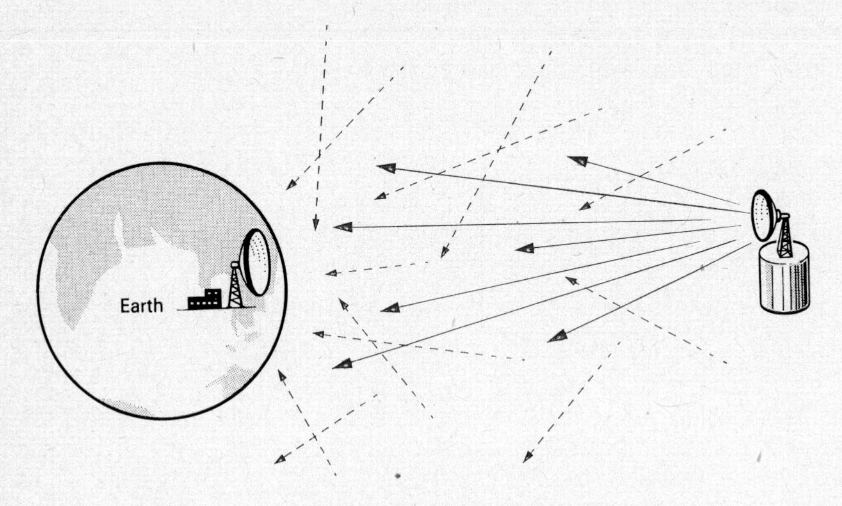

Figure 7 Satellite signal and stellar noise reaching Earth

The problem could have been solved by providing the Mariner probe with more powerful transmitters or a larger transmitting aerial. However, both of these solutions would have increased the mass of the space probe, and, since 1 kg increase in probe mass requires at least 1000 kg increase in launcher mass, the extra cost incurred prohibits the use of either of these solutions. Therefore a solution to the problem which did not require extra mass was favoured. It is just such a solution that a study of the properties of random signals makes possible.

4.2 The evoked response of the human nervous system

The evoked response of neural networks (often called the electrical pathways of the body) is the response produced by a stimulus.

Studies of such evoked responses due to a given stimulus are, however, always confused by the fact that nerves in the network are in a state of continuous activity. The continuous activity, although presumably meaningful within the nervous system itself, is apparently random when observed from outside. Thus the evoked response caused by the externally applied electrical impulse is disturbed by the continuous, background activity of the nerves. (See Figure 8.)

This noise component is always present together with the required signal. We shall see later how to separate the two.

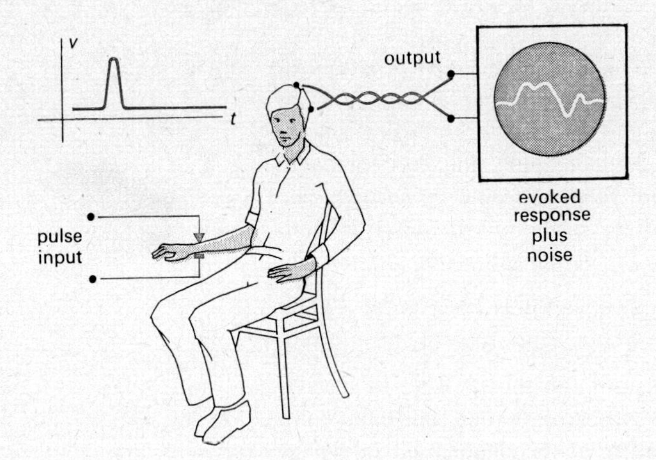

Figure 8 The experimental arrangement for observing the evoked response of the human nervous system

4.3 Measurement of the velocity flow rate of sewage

In order to increase the use of automation of sewage treatment it is necessary to have a flow meter that can measure flow rates in inhomogeneous liquids such as sewage. Orifice type flow meters, which were described in Units 8/9/10, tend to block when measuring dirty liquids, so a new technique for making flow measurements on such liquids has been developed.

The essence of the problem is to find some way of timing the flow of a given section of liquid between two points. This is done in the new method by continuously measuring the conductivity of the liquid. When a particular profile of conductivity appears first at one point and then at the other the time difference between these two instants gives you the flow rate. The trouble is that the liquid swirls and eddies as it flows, introducing 'noise' into the measurement. The variation of conductivity is not the same at the two measurement points. The measurement technique used, and described in Unit 13, separates this noise component from the wanted information.

Quantifying a waveform

The task of detecting the wanted signal in the presence of interference depends upon making use of a knowledge of the characteristics of both. Using an electrical filter to remove high-frequency noise from a low-frequency signal is an example of one way of doing this; but for other types of signal or interference a different strategy would be needed. The first step towards selecting an appropriate strategy is to describe both signal and noise adequately. How to do this is the subject of this section.

Now, as already explained, a full description can only be given of a finite section of a signal. The properties of the whole signal can only be inferred.

The basic problem, therefore, is to measure, and thus be able to specify, the waveform over a time interval T_o, the observation time. The bandwidth of the signal is limited either by the nature of the signal itself or by the method of observation. The specification is to be made on the basis of the properties of the waveform itself and is not to rely on any other information except a knowledge of this bandwidth. This is what is meant by quantifying the waveform.

One way of quantifying the waveform is to 'specify' the Fourier coefficients of the waveform, that is, the magnitude and phase of the frequency components of the waveform.

Another way is to record its instantaneous amplitude at a series of instants throughout the duration of the waveform.

Figure 9 shows two waveforms together with the magnitude and sign of a series of samples shown in red. These sample values are tabulated in Table 1, and for each waveform they provide a form of quantitative

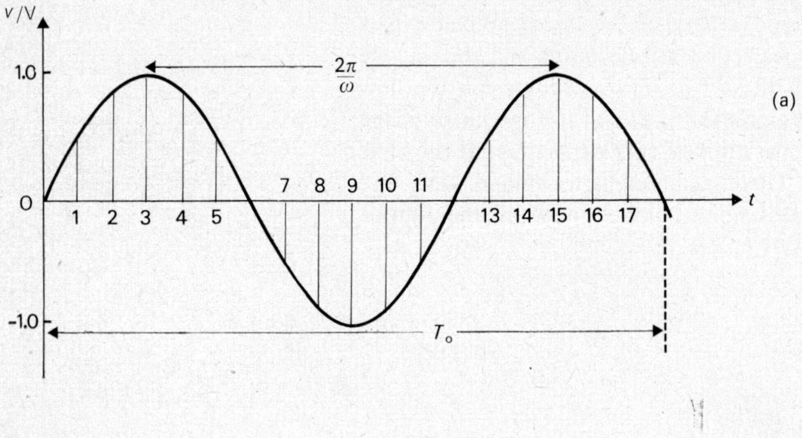

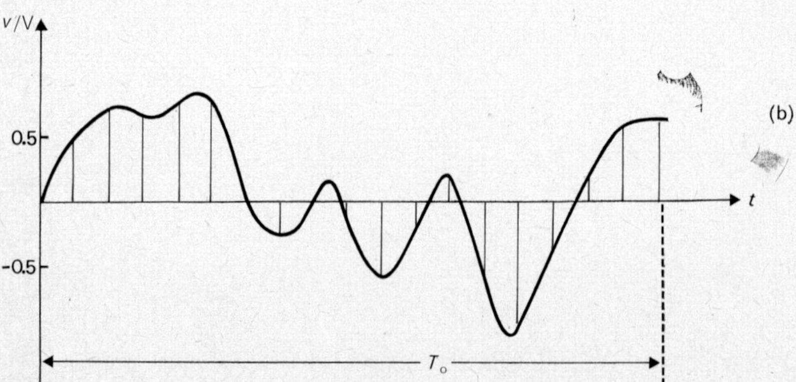

Figure 9 The sample values of (a) a sinusoid and (b) a random signal

Table 1

Sample number	Sample values A	Sample values B
1	0.5	0.5
2	0.87	0.75
3	1.0	0.70
4	0.87	0.78
5	0.5	0.80
6	0	0
7	−0.5	−0.25
8	−0.87	0
9	−1.0	−0.15
10	−0.87	−0.6
11	−0.5	−0.2
12	0	0.15
13	0.5	−0.67
14	0.87	−1.1
15	1.0	−0.35
16	0.87	0.2
17	0.5	0.6
18	0	0.65

description. It is necessary, however, to determine how many samples are needed to quantify the waveform fully. As we shall see, this number depends upon the bandwidth of the waveform within the observation time T_o.

The number of samples needed is specified by the sampling theorem, which I shall state, explain and demonstrate, but which I shall not attempt to prove mathematically.

5.1 The sampling theorem

The sampling theorem states that:

A continuous signal can be represented completely by, and reconstructed from, a set of instantaneous measurements or samples of its voltage which are made at equally spaced times. The interval between such samples must be less than one-half the period of the highest-frequency component in the signal.

To understand what this means in relation to a sine wave, look at Figure 10. It shows a series of sine waves of different frequencies, each of which has been sampled at the same instants of time. In the first two cases the frequency of the signal is less than half the reciprocal of the time between the samples. That is, f is less than $1/(2T)$. There is only one sinusoid that can be drawn through the sample values in each case. It is shown by the solid black line in Figures 10(a) and (b).

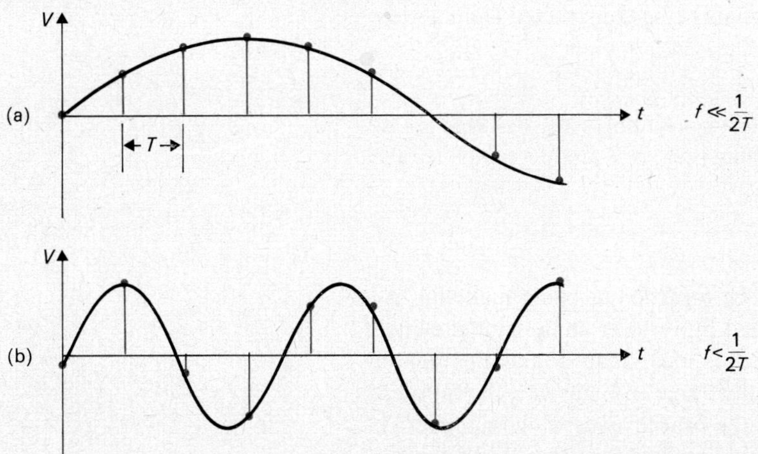

Figure 10(a) and (b) The sampling of sinusoids whose frequencies are less than $1/(2T)$

Figures 10(c) and (d) show two special cases in which the sampling rate is exactly twice the frequency of the sine wave. In the case of Figure 10(c) it is clear that each sample has zero value and consequently provides no information about the sine wave. By shifting the sample instants somewhat in time, as shown in Figure 10(d), it is apparent that the samples now fully describe the frequency of the sine wave, but not its amplitude. Thus here the samples do *not* describe the waveform adequately.

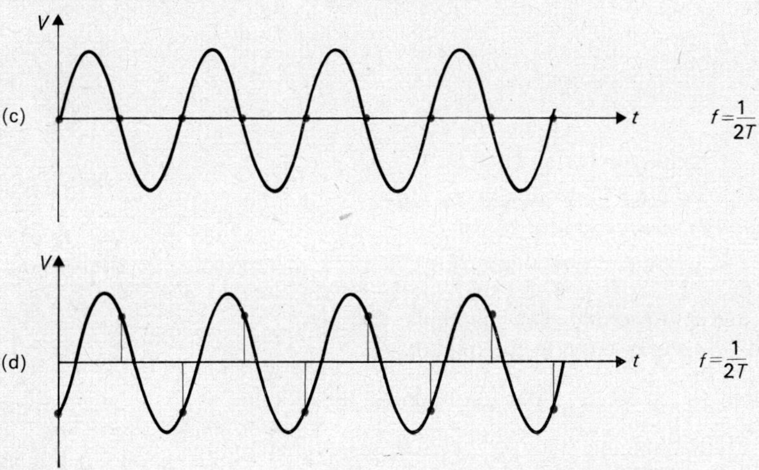

Figure 10(c) and (d) The sampling of sinusoids whose frequencies are equal to $1/(2T)$

Figures 10(e) and (f) show cases in which the sampling rate is too low in relation to the frequency of the sinusoid. The problem that arises in such cases as these is illustrated by the broken red sinusoidal curves. These red curves show sine waves of a lower frequency which nevertheless pass through the same sample points as those obtained from the original signal. This being so, it is evident that, given only the values of the samples, it is not possible to deduce which of the two sine waves shown was the original signal. Thus, *to avoid ambiguity, it is necessary to sample the sine wave at a rate corresponding to more than twice the frequency of the sine wave.*

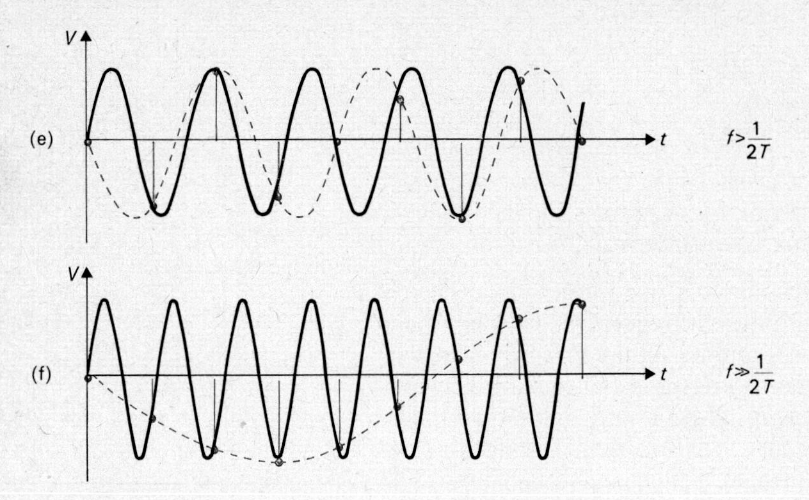

Figure 10(e) and (f) The sampling of sinusoids whose frequencies are greater than 1/(2T). Aliasing is indicated by the dashed lines

This phenomenon of two sinusoidal signals being constructed from a given set of sample values is called *aliasing*.

aliasing

If the frequency of the sinusoid in Figure 10(b) is 285 Hz, what are the approximate values of the sampling interval T and the sampling rate?

1 ms, 1 kHz.

Now, if we accept that any signal can be regarded as being made up of a spectrum of sinusoidal waves, as implied by Fourier analysis, it is evident that, if we choose a sampling rate which is more than double the *highest-frequency* component in the signal, as illustrated in Figure 11, the sampling rate will be high enough to specify *all* the other frequency components of the signal. It should therefore be possible to specify such a 'band-limited' signal without ambiguity or error, from the samples taken at a rate of $1/T > 2f_c$. This is indicated in Figure 11.

Physically what this means is this, that if correct interpolation between samples a time T apart is to be possible, the sampling rate should be more than twice the highest frequency in the signal.

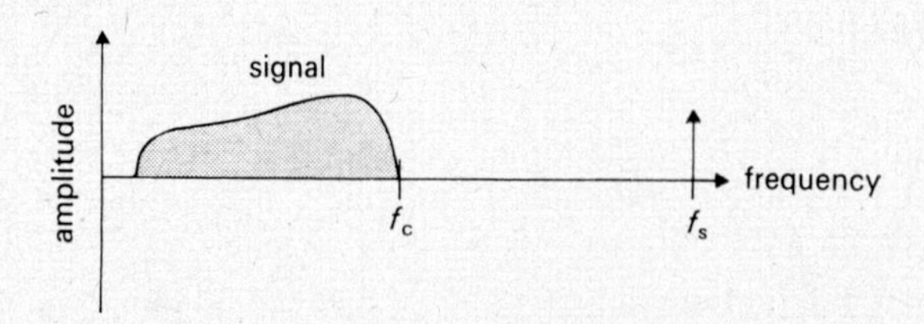

Figure 11 The shaded area shows the spectrum of a signal to be sampled. Its highest-frequency component is f_c. The sampling frequency f_s must be at least $2f_c$

Figure 12 shows a waveform and its spectrum. Do you think the waveform can be fully specified by samples taken at the frequency f_s shown?

No, the sampling rate is too low.

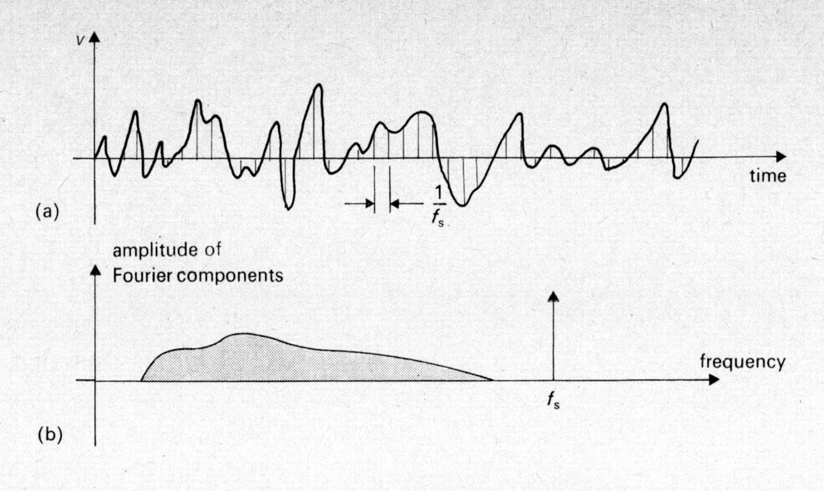

Figure 12 A signal spectrum and a suggested sampling frequency f_s

Now let us return to the question we were considering before we embarked upon the sampling theorem. I posed the question, 'How many samples are required to ensure that a segment of signal of duration T_o is fully specified by the magnitudes and signs of these samples?'

We can now answer the question.

First we must know the bandwidth of the signal. For the time being I shall speak only of signals whose spectra extend from zero up to some cut-off frequency f_c. The bandwidth of a signal will be determined by the combined bandwidth of its transducer and an appropriately chosen filter. The effective cut-off frequency of the *signal*, f_c, must be decided in relation to the required accuracy. For instance, suppose the set of samples is required to represent the waveform with an accuracy of 1 per cent. In that case, the spectrum of the signal is considered to extend up to a cut-off frequency f_c, above which the frequency components of the signal contribute less than 1 per cent to the waveform. Note that, as a consequence of this, what is considered to be the cut-off frequency of the *signal*, f_c, is commonly much higher than the -3 dB cut-off frequency of the combined response of the transducer and the filter which is used before sampling.

What is the minimum sampling rate in order to specify such a signal?

$2f_c$.

Then we must multiply this rate by the observation time T_o.

Thus a segment of such a signal of duration T_o *can be specified by* $2f_cT_o$ *samples.*

If the frequency range of the spectrum is not limited in some way, the signal cannot be quantified.

Note, however, as stated earlier, this method of specifying a time-limited, band-limited waveform in terms of $2f_cT_o$ samples is not the only method. Such a waveform can equally be specified in terms of Fourier components, and again it is the case that the number of Fourier coefficients needed is $2f_cT_o$. Thus a more complete statement is that a band-limited waveform of duration T_o can be specified in terms of $2f_cT_o$ *quantities.*

How can you specify the Fourier components of a non-periodic waveform of finite duration?

This question arises because, as you should recall, Fourier series, and therefore Fourier components, describe a periodic signal. But now we have a waveform with no periodicity but finite duration.

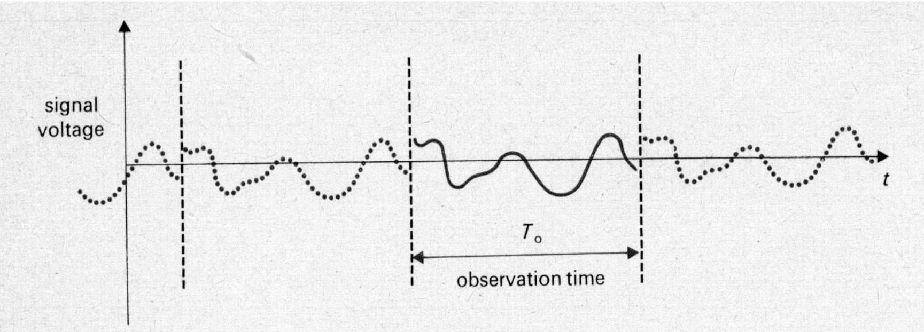

Figure 13 The periodic signal formed by treating an observed waveform as one period of a periodic signal

In order to specify the waveform using Fourier components, we have to specify a set of sinusoids, both in amplitude and phase, whose sum reproduces the given waveform within the observation time T_o. There is no limit to the number of sets of sinusoids which will do this but the smallest set, or minimum set, is that which generates a periodic signal, of period $1/T_o$, of which one period is the observed waveform, as indicated in Figure 13. As explained in the answer to SAQ 2, this set again requires $2f_cT_o$ quantities to specify it.

SAQ 2

SAQ 2

Set up the general Fourier series for such a signal. How many independent coefficients does this series contain?

5.2 An implication of the sampling theorem

The sampling theorem tells us that a continuous signal can be described completely by an endless series of sample values, provided the samples are separated by a small enough interval, less than $1/(2f_c)$. An implication of this is that, if a signal is going to be transmitted from place to place, it is unnecessary to send a continuous representation of the signal. All that need be transmitted are the sample values. In fact, as you will see in later units, it is often advantageous to transmit signals in this way.

However, it may be that the received signal must be converted back to its original form, before, for example, being fed into a chart recorder. The sampling theorem states that such a reconstitution of the original signal is possible. But how is it to be done in practice?

The technique is simply to pass the sequence of samples through a low-pass filter of a particular cut-off frequency, as indicated in Figure 14.

As we noted earlier, a low-pass filter is a circuit which attenuates any frequency components of a signal which are above a particular frequency, termed the 'cut-off frequency', more than it attenuates any frequency components below the cut-off frequency. Before going on to discuss how it is used to recover a signal from its samples, let us consider the effect of such a circuit upon a square wave like that of Figure 15(a).

What frequency components has such a square wave?

You should recall from Units 8/9/10 that the Fourier components of such a waveform are the fundamental frequency of the waveform together

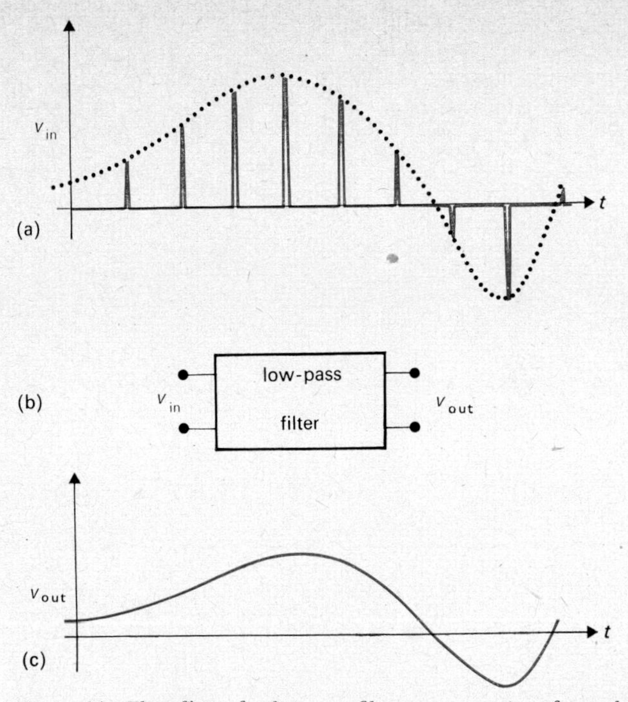

Figure 14 The effect of a low-pass filter upon a series of samples. The sampling rate in (a) is greater than $2f_c$. The cut-off frequency of the low-pass filter in (b) is half the sample frequency. The output shown in (c) is reduced in amplitude as compared with the sample amplitudes, is somewhat delayed, and, if the sample frequency is only a little greater than $2f_c$, may suffer from phase distortion

with its odd harmonics in decreasing amplitude, as shown in Figure 15(b). If this signal is applied to the input terminals of a low-pass filter whose cut-off frequency is, say, four times the fundamental frequency, as indicated in Figure 15(c), then the resulting amplitude spectrum will be as shown in Figure 15(d).

What will the output waveform of the filter be like?

Did you decide this question is unanswerable? You should have done. While it is true that the output waveform consists of the sum of the first and third harmonics, as shown in Figure 15(d) (since the filter attenuates higher harmonics), it is not possible to state the form of the resulting waveform without knowing the *phase* of the third harmonic relative to the fundamental.

Figures 15(e) and (f) show two possible outputs corresponding to different relative phases of the first and third harmonics. Evidently, therefore, both the phase and magnitude response of the filter must be properly controlled. It turns out that the simplest way to avoid phase distortion due to the filter is to design the system so that the highest signal frequency is far below the filter cut-off frequency. As we shall see, this is not difficult to achieve when reconstructing a signal from a sequence of samples.

However, I do not want you to consider filters in detail in this course, though the subject is touched upon again in Unit 12. Apart from this important point about phase shift in filters, the thing to notice about the effect of the filter upon the waveform is how the sharp steps in the input voltage waveform are smoothed out by the filter. The Fourier components needed to produce sharp steps are removed by the filter.

If the low-pass filter had a cut-off frequency of *twice* the fundamental frequency of the square wave, what form would the output waveform take?

Since the filter now only passes the fundamental (there is no second harmonic in the spectrum of a square wave), the output waveform must be a sinusoid. As a result of the phase shift in the filter, however, the sinusoid will be delayed to some extent as compared with the input fundamental component.

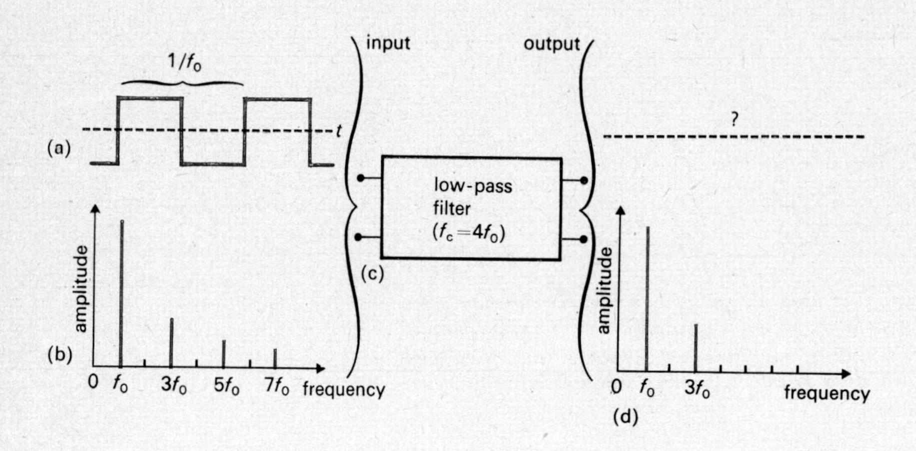

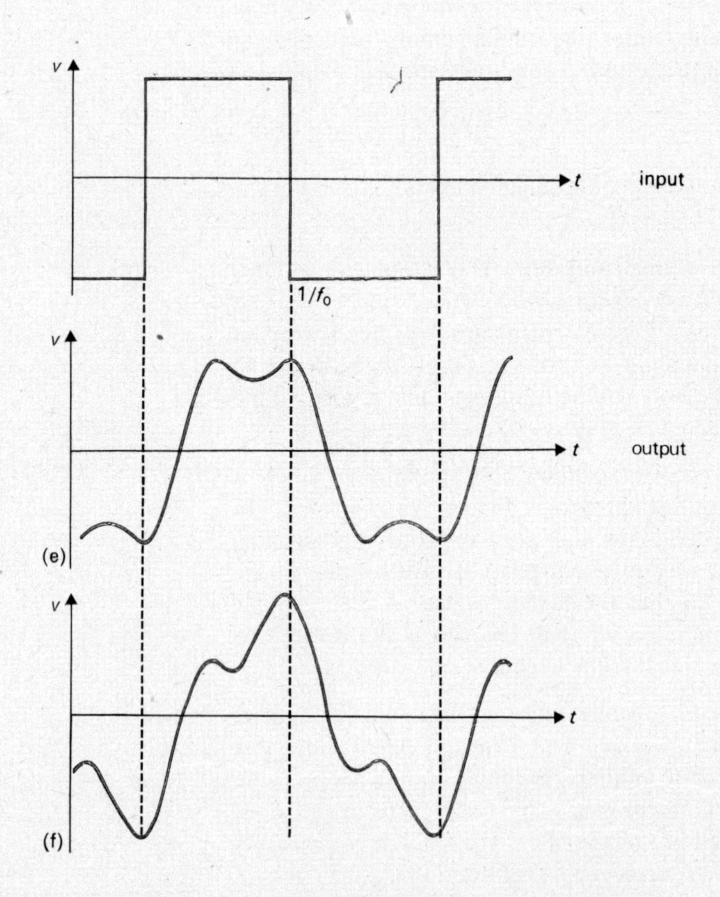

Figure 15 The effect of a low-pass filter with cut-off frequency $4f_0$ upon a square wave of fundamental frequency f_0: (a) and (b) show the input waveform and spectrum; (c) represents the filter; (d) shows the spectrum of the output; (e) shows the output from a linear phase shift filter; (f) shows the output when the third harmonic is delayed more than the fundamental component of the waveform

Now let us return to a consideration of the effect of a low-pass filter upon a series of samples of a signal, obtained at a rate of more than $2f_c$. Let us suppose the filter cut-off frequency is half the sampling frequency, so it is more than f_c.

It is possible to choose a sampling frequency only just higher than $2f_c$ and construct a filter of cut-off frequency just above f_c (Figure 16a). But it would have to have a sharp cut-off and its phase response would lead to considerable phase distortion of the signal. In general, therefore, a sampling frequency much higher than $2f_c$ is chosen, so that the filter cut-off frequency is also well above f_c. This ensures that the final output does not experience significant attenuation or phase distortion. See Figure 16(b).

The fundamental sampling frequency is well above the filter cut-off frequency, so the filter can be designed to attenuate it strongly. But all the frequency components of the original signal are contained in these equally spaced samples and these will not be attenuated significantly by the filter. Thus the output of the filter is a replica of the input waveform.

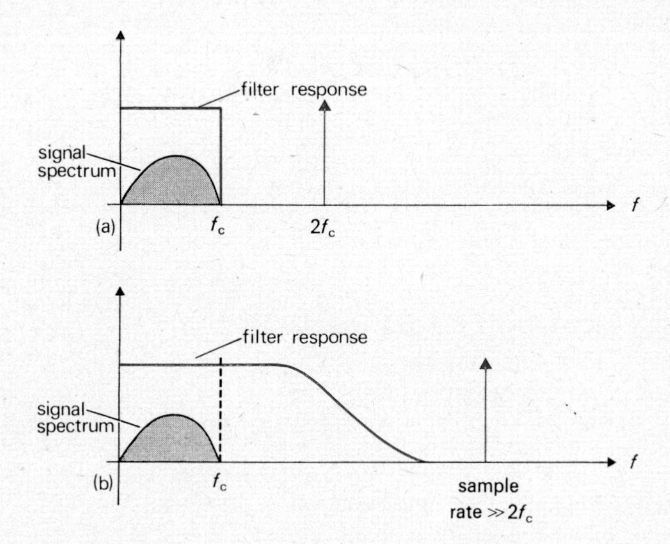

Figure 16 Low-pass filter responses: (a) an idealized but unrealistic filter; (b) a more practical response suitable for use with a sample rate greater than $2f_c$. *(Such a response is what you would obtain with a practical filter if you plotted against their frequency the ratio of the amplitudes of the output and input signals passing through the filter. Such test signals, of course, are sinusoidal.) The shaded areas represent a possible signal spectrum*

This discussion of the implications of the sampling theorem is simply to emphasize the validity of the sampling theorem as a basis for quantifying a waveform. From now on therefore I shall take it as fully accepted that:

A band-limited waveform of bandwidth f_c and duration T_o can be fully specified by $2f_cT_o$ quantities, within the accuracy of the samples.

5.3 Random and deterministic waveforms

It is sometimes of interest to attempt to describe or define what is meant by a 'random' waveform and to distinguish it from deterministic or periodic waveforms. An understanding of these distinctions does help you to avoid unnecessary errors.

random waveform

Briefly, a band-limited, time-limited, *random waveform* is one which cannot be specified with fewer than $2f_cT_o$ quantities (where, as before, f_c is the bandwidth and T_o is the observation time or duration of the waveform). *Deterministic waveforms* can be specified with fewer than $2f_cT_o$ quantities.

deterministic waveform

The two waveforms of Figure 9 fall into different categories. Figure 9(a) is a sinusoidal waveform and can therefore be specified by the function

$$v = A\sin(\omega t+\varphi) \quad (0<t<T_o).*$$

The voltage as a function of time can be specified by three quantities, the amplitude A, the frequency ω and the phase φ. These three quantities enable any set of instantaneous values within the observation time to be calculated, so they fully specify the waveform.

The waveform of Figure 9(b) may or may not be capable of being represented by a mathematical function containing fewer than $2f_cT_o$ variables. If we do not know of any such function, even if there is one, we must in practice treat the waveform as random, that is, we must regard it as one which cannot be specified by fewer than $2f_cT_o$ quantities.

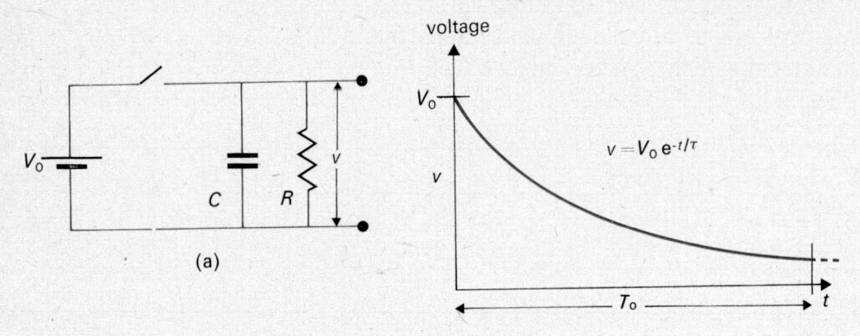

Figure 17 A non-periodic deterministic waveform. The switch in the circuit is opened at $t=0$ *and the output voltage* v *then falls exponentially* ($\tau=CR$)

A voltage which has the form of a dying exponential – like the voltage across a discharging capacitor – is another kind of deterministic waveform (Figure 17). The voltage as a function of time can be expressed by the equation

$$v = V_0 e^{-t/\tau} \quad (0<t<T_o).$$

Here only parameters V_0 and $1/\tau$ are needed to express the time variation of voltage.

Notice that, whereas the sinusoid was deterministic and periodic, the dying exponential is deterministic but *aperiodic* – it does not contain a recurring pattern.

Now, the main purpose of trying to draw a distinction between random and deterministic waveforms is to warn you against a misconception which can easily arise from our considerations of random signals in the next few sections. It is that interference is random whereas the required signal is deterministic. According to such a view the detection of a signal comprises the separations of the random component (the noise) from the required deterministic waveform. Such a view is quite false. You can see that it must be false as follows.

The wanted signal is 'wanted' because of the information it carries. If it is a deterministic waveform, like a sine wave, the amount of information it can convey is limited by the number of quantities involved in specifying the waveform – only three for a continuous sine wave.

The most information a waveform can carry is dependent upon the number of independent quantities needed to specify it. Thus a waveform which carries the most possible information cannot be specified by fewer than $2f_cT_o$ quantities. But this makes it a random waveform.

*$0<t<T_o$ *means that in this case* t *has values between* 0 *and* T_o.

Thus information-carrying signals must also be, to some extent at least, random ones. If their values as a function of time can be calculated, they cannot be conveying new information. As a consequence, I hope you can see that, in general, we cannot expect to find a universal strategy for separating a wanted signal from noise, for they may well be indistinguishable.

Similarly, it is equally false to suppose that interfering signals are necessarily random, though many are. They may well be deterministic, like 50 Hz interference from the a.c. mains supply.

5.4 Summary

1 The sampling theorem states that:

A continuous signal can be represented completely by, and reconstructed from, a set of instantaneous samples of its voltage which are taken at equally spaced times. The interval between such samples must be less than half the period of the highest-frequency component in the signal.

2 A waveform of duration T_o can be specified by $2f_cT_o$ quantities, where f_c is the bandwidth of the waveform. This applies to the case of a waveform whose spectrum extends from zero frequency to an upper limit f_c. The degree of attenuation needed at f_c depends on the accuracy required.

The following comments apply to such signals.

3 The quantities used to specify a waveform may be instantaneous sample values or Fourier components. (Other sets of quantities are also possible, but do not concern us here.)

4 A sequence of equally spaced samples of a waveform can be used to regenerate the waveform. The method is to pass the samples through a low-pass filter. The cut-off frequency of the low-pass filter must be more than f_c and less than half the sampling frequency. Its phase response should be sufficiently linear to achieve the accuracy required in the reconstituted waveform. (In practice a sampling frequency much greater than $2f_c$ is normally used. This eases the design of the low-pass filter.)

5 Waveforms which can be specified with fewer than $2f_cT_o$ quantities are, to some degree, deterministic. Some deterministic waveforms are aperiodic, some are periodic.

6 Random waveforms cannot be specified with fewer than $2f_cT_o$ quantities.

7 Wanted signals are usually random, because such signals can carry more information than deterministic ones.

8 Interference can be either random or deterministic.

Descriptions of random waveforms

We now know how to quantify a waveform, either by samples or by Fourier components. But a list of numbers, such as those of Table 1, is not a very convenient way of describing the nature of a whole waveform. There are a number of types of average of a set of quantities describing a waveform which are found to be useful as a means of distinguishing one kind of random waveform from another. This section is concerned with specifying and explaining some of them.

Remember, in every case we can only deal with a finite segment of a signal: the waveform within the observation time. It is a segment of a signal which we can hold onto, so to speak. That is, we can list and remember its sample values, and perform calculation upon them. We do not have to calculate or observe these averages in 'real time' as the signal arrives. We can allow a computer to perform its calculation upon the sample values of the waveform at any convenient time. But we are always limited in practice to a finite duration of the signal.

The first two parameters or characteristics we shall be concerned with are:

1 the mean value of a waveform;

2 the root mean square value of a waveform;

3 the probability density function of a signal.

These parameters are all concerned with different ways of handling the instantaneous values of a signal and are not in any way dependent upon the range of frequencies contained within the signal.

The next two parameters are concerned with the range of frequencies contained in the signal. They are:

4 the power density spectrum of the signal;

5 the autocorrelation function of the signal.

Parameters 1–3 are discussed in this unit. Parameters 4 and 5 are considered in Unit 12.

Since each one of these five parameters, or characteristics, is an average of some kind, its value or properties, for any particular waveform, depend to some extent upon the observation time and even upon which section of the signal forms the waveform being considered. An average value taken over a waveform of one particular period will differ a little from the average value taken over double that period, or from an average taken over a different similar period. A random signal is continuously fluctuating, so that, even if nothing changes in the manner in which the signal is generated, there can be no guarantee that an average value obtained from one portion of the signal will be exactly the same as that obtained from another portion. Indeed, if the averages were exactly the same, it would be exceptional. In fact, averages obtained for a whole series of waveforms taken from the signal will themselves have a random distribution.

Suppose you are repeatedly observing a random signal and calculating its mean value* over a certain observation time. You find that the values you calculate fluctuate randomly. How do you know whether this is due, for example, to the presence of a wanted signal to which noise has been

**See section 6.1 if you do not already know what this means.*

added, or is due solely to random fluctuation of the random signal? The answer is that you cannot know for sure.

But if the mean values you calculate (as well as some of the other averages I shall be describing) themselves show only random fluctuation about a more or less stationary mean, the observed random signal is said to be *stationary*. Even this 'stationary' mean will be different each time you calculate it – so you can never be quite sure, although you can state the degree of your confidence (see Section 7). Thus a random signal which gives no statistical grounds for judging that it has changed in character is called a *stationary* random signal. It is an easy idea which is very difficult to express precisely. We shall always be assuming that noise which interferes with our signals is stationary in this sense.

The six kinds of averages listed can be obtained for periodic waveforms as well as for random ones. The procedures needed to obtain them are identical but a good deal easier to describe and explain using periodic waveforms. So I shall consider each of these kinds of average with reference to simple periodic waveforms like square waves and sine waves, and then later consider random waveforms.

We shall only consider waveforms as specified by $2f_cT_o$ *sample* values. We shall not be considering averages of Fourier coefficients. Deterministic waveforms can, of course, be specified with fewer data.

6.1 The mean value of a waveform

The mean, or average, value of a waveform is the arithmetic mean of the sampled data describing it. The calculation of an arithmetic mean should need no further explanation. (See answer to SAQ 3 if you do not know how to work out an average.)

SAQ 3

SAQ 3

Calculate the mean value of the sinusoidal waveform and of the random waveform whose samples are tabulated in Table 1. What rule applies to inferring the average of a periodic *signal*, given only a segment of it?

Why is the mean value of the random waveform not zero?

6.2 The r.m.s. value of a waveform

The r.m.s. value of a time-varying voltage waveform of duration T_o is equal to the constant, direct voltage which will dissipate in time T_o in a given resistor the same energy as the varying voltage.

'R.M.S.' means 'root mean square' and is the square root of the mean of the square of the waveform. Since the power dissipated by a resistor at any instant is proportional to the square of the voltage applied to it, these two expressions of the meaning of r.m.s. can be seen to be saying the same thing.

If the waveform whose r.m.s. voltage is required has been quantified by sampling, it can be shown that the r.m.s. voltage is simply the square root of the sum of the squares of all the sample values. That is, to obtain the r.m.s. voltage you:

1 square all the sample values;
2 add them all together;
3 divide by the number of samples;
4 take the square root.

Mathematically, the r.m.s. value of n sample values $v_1, v_2, v_3, v_4, \ldots, v_n$ of a signal may be written

$$v_{rms}=\sqrt{\frac{v_1^2+v_2^2+v_3^2+v_4^2+\ldots+v_n^2}{n}}. \quad (1)$$

As with taking the mean, it is necessary to consider only integral numbers of whole periods of a periodic wave if you wish to infer the r.m.s. voltage of a continuous periodic signal.

SAQ 4

SAQ 4

Given the sample values tabulated in Table 1, what r.m.s. values would you infer for the sinusoidal signal, and for the noise signal from which the waveforms of Figure 9 were derived?

6.3 Comment

Ordinary, moving-coil, d.c. meters do not indicate instantaneous values of the voltage applied to them. Due to the inertia of the mechanical parts they indicate average voltages. The effective observation time for such a meter depends upon the *response* time of the mechanical movement. Light-weight movements will just show a response to the 50 Hz mains fiequency – so the effective observation time is of the order of $\frac{1}{50}$ s, although there is no precise period specifiable.

Electronic meters which indicate r.m.s. voltages can also be obtained. They too have a response time which in effect defines their observation time. Some r.m.s. voltmeters are equipped with switches to select the averaging time. Thus their observation time is adjustable.

Either sort of meter used to measure the mean or r.m.s. value of a random signal will show a fluctuating value.

The pointer of the meter will not stay at one particular value (unless the signal is filtered to remove frequency components whose period exceeds the meter averaging time).

Why is this?

It is because the set of values within the observation time to which the instrument automatically responds, fluctuates. The average for one $\frac{1}{50}$ s, say, differs from that of another, so the indicated value fluctuates.

In general, the longer the averaging time, the smaller the fluctuations around the mean, for reasons which will be explained shortly. But it is important to realize that random waveforms also have somewhat random averages.

6.4 The standard deviation of a waveform

The standard deviation of a time-varying voltage is a measure of the range of departure of the voltage from its mean value.

The standard deviation, like the mean value, is calculated using sample values. To obtain the standard deviation you:

1 subtract the mean value from each of the sample values;

2 square the results;

3 add them all together;

4 divide by the number of samples;

5 take the square root.

Mathematically, the standard deviation σ of n sample values v_1, v_2, v_3, ..., v_n, with a mean value v_m, may be written as

$$\sigma = \sqrt{\frac{(v_1 - v_m)^2 + (v_2 - v_m)^2 + \ldots + (v_n - v_m)^2}{n}}. \quad (2)$$

Suppose the mean value of the sample values is zero, what then does equation (2) reduce to?

Equation (2) becomes the same as equation (1) for calculating the r.m.s. value of a set of samples. In fact, the r.m.s. value of a waveform with zero mean is equal to the standard deviation of the waveform.

6.5 The probability density function of a signal

The probability density function of a signal is most easily described by means of a graph. It represents the likelihood of sample values of the signal falling within a specified range of values.

In order to explain this idea I shall first consider a rather different example. Suppose you measured the heights of all the men in a particular town during one day (the observation time), and evaluated the fraction of the total number of men whose heights were within one-inch ranges from 4′ 8″ to 6′ 4″. You could then draw a *histogram* of your results, as in Figure 18(a). This histogram is usually called a *frequency distribution* by statisticians, but because of the different meaning given to 'frequency' in instrumentation I shall use the word 'histogram'.

histogram

frequency distribution

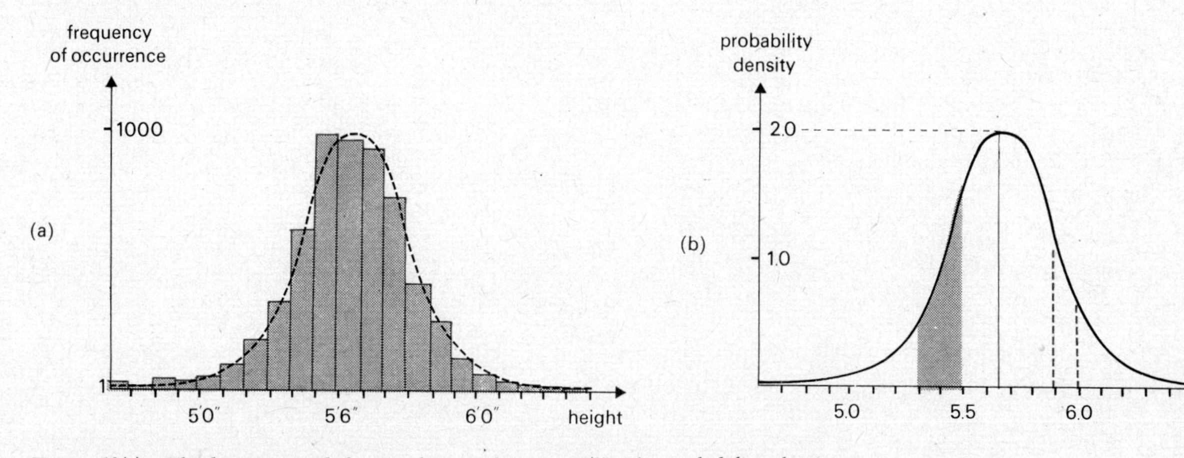

Figure 18(a) The histogram of the population of a town. (b) The probability density function derived from it (and shown dotted in a). Probability is represented by the area under the graph. The total area must have the value 1.0, so that the peak of the graph is at a probability density of approximately 2.0. The shaded area is the probability of finding a man within the limits of 5.3 and 5.5 feet

If you were to do it all again on a different day, you would probably obtain a different, but similar, histogram.

Histograms represent the results of measurements – just like experimental points on a graph. Often when you have plotted these points on a graph, you draw a smooth curve through them to represent the underlying property or characteristic revealed by these points – though with some error. In the same way, if we now draw a smooth curve to show the envelope of the histogram, we obtain what is called the probability density function. It represents the likelihood of your finding particular heights or ranges of heights next time you try. The probability density in any range is the number in that range divided by the total number.

This probability density function is reproduced in Figure 18(b). Notice that, while the histogram has clearly marked ranges of heights, the density function does not. So how should you read values of likelihood from your density function? From the *area* under the graph. The probability of finding a man with height between 5.6254 ft and 5.6255 ft is very small, even though this is the height at which the function is a maximum. The area of the narrow strip between these heights represents the probability of finding such a man, and it is very small, The probability of finding a man with height between 5.3 and 5.5 ft is represented by the area of the broader shaded strip; it is about 0.2. If all the men in the town were within the limits of the function shown (4.5 ft and 6.5 ft), it is *certain* that the men you observe are within this range. When a probability is a certainty it is given the value 1.0. Thus *the total area under a graph of a probability density function is* 1.0.

What is your estimate of the probability of finding a man between 5.9 and 6.0 ft in height?

You have to estimate the area of the strip marked as a portion of the whole area under the graph. In this case it is approximately $\frac{1}{12}$, so the probability is 0.08.

Having given this introduction, I shall now consider some more examples, but this time related to sampled waveforms.

6.5.1 *A sawtooth waveform*

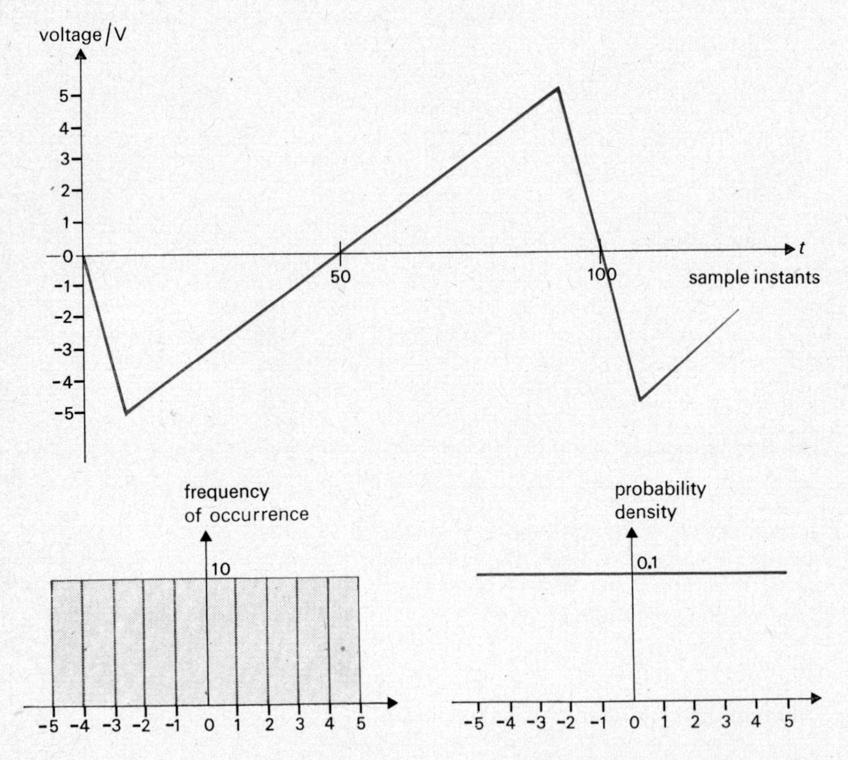

Figure 19(a) A sawtooth waveform sampled 100 *times. (b) The histogram of the voltage values of the waveform. (c) The derived probability density function*

Figure 19(a) shows a sawtooth, or zig-zag, waveform. The voltage increases linearly from its minimum, negative value to its maximum value and then drops more rapidly back to its minimum value again before once more commencing to rise. Now, if we take samples of this signal at frequent intervals during the linear rise of the voltage, it is evident that the frequency of occurrence of any particular instantaneous amplitude is the same as that of any other. To be more precise, the proportion of the time that the instantaneous voltage falls within a given range of possible voltage is the same for all equally sized ranges of voltage. This is a direct result of the fact that the waveform consists of straight lines.

In Figure 19(a) ten ranges of voltage are marked on the ordinate (from −5 V to −4 V, from −4 V to −3 V, etc.). If the abscissa, or the time axis, is divided up into 100 equal sampling intervals during one cycle of the sawtooth waveform, it is evident that there will be ten samples within each of the specified voltage ranges. Thus the histogram we obtain is as shown in Figure 19(b). Here there is no randomness in the waveform, so the probability density function for the sawtooth waveform is simply as shown in Figure 19(c). It is a straight line at a probability density of $\frac{1}{10}$. This value is obtained by dividing the frequency of occurrence (ten occasions) by the total number of samples, 100 in this case.

From Figure 19(c) you can see that the value of the probability density function corresponding to a voltage of +1 V is equal to $\frac{1}{10}$. Does this mean that the chances of obtaining a sample value of *exactly* +1 V is one in ten?

No! Probability *density* is the probability of occurrence of a range of values of a variable, *divided by that range*. Or, in other words, as I stated earlier, the *probability* is the area under the graph. If the probability density is constant over the range,

Probability = probability density × range.

Thus, if, in the example illustrated by Figure 19, you want to know the probability of obtaining a value between −1 V and +3 V, you multiply the value of the probability density function by the size of the range. In this example the value of the probability density function is constant (in general it will vary over a range) and equal to $\frac{1}{10}$. The range from −1 V to +3 V is equal to 4 V. Therefore the probability of getting a value between −1 V and +3 V is equal to $\frac{1}{10} \times 4 = \frac{4}{10}$. This means that for a sample taken at a random instant there is a chance of two in five of obtaining a value between −1 V and +3 V

This is merely an evaluation of the area under the graph of the probability density function between −1 V and +3 V.

Notice again that the area under the curve of probability density function must equal 1, because it is certain that all sample values of the waveform lie between the limits of ±5 V. This is another way of establishing that the value on the probability density axis is $\frac{1}{10}$. Indeed, it is the method that is usually used.

Notice that the units of probability density in this example are reciprocal volts (V^{-1}), since probability is a number.

Of course, probability density functions are normally used for non-deterministic waveforms because they refer to probabilities rather than to certainties. In the case of this idealized sawtooth waveform the number of instantaneous amplitudes within the specified ranges is not a matter of probability. We are in no doubt that for 100 samples there are ten instantaneous amplitudes within each voltage range.

So whether you are concerned with a random waveform or a deterministic one you can record your results in the form of a histogram. This displays your data and is not a matter of probability. You can then draw a curve through your data and obtain for each an estimate of a probability density function. The probability aspect of the random waveform, however, only emerges when you repeat these processes on the same signals. With the deterministic waveforms your probability density functions will always turn out to be the same. With the random waveforms each set of data and each derived density function will differ somewhat from every other one.

In practice some noise, however small, must be interfering with the observation of any waveform, so that the signal will indeed be partly a random one and the sample values will be partly probabilistic – not absolute certainties.

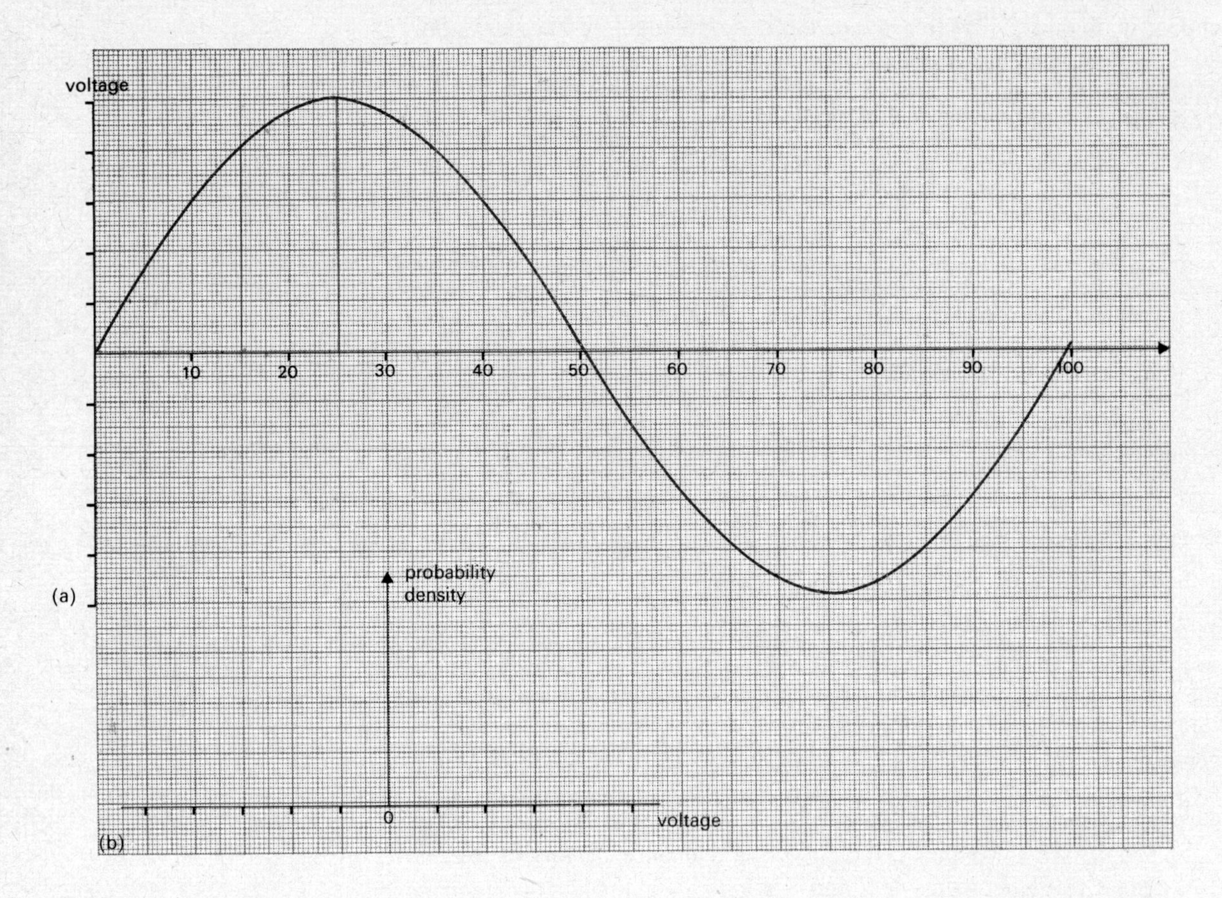

Figure 20(a) A sinusoidal wave with 100 *sample instants. (b) The axes for a plot of the probability density function*

SAQ 5

SAQ 5

Figure 20(a) shows one cycle of a sine wave with the time axis marked with 100 sample instants. Divide the peak-to-peak amplitude of the sine wave into ten equal ranges and count the number of sample values falling within each range. Plot the histogram of your results and estimate the probability density function of the sine wave on the graph paper provided. Use the axes as shown in Figure 20(b).

6.5.2 *A portion of a random signal*

Figure 21(a) shows an example of a random waveform. It is a segment of what is called Gaussian noise. The peak-to-peak range of the samples of the wave shown is divided into ten equal ranges and the time axis is divided into fifty equal intervals. If you were to go through the same procedure as you used to estimate the probability density function of your sine wave, you should first obtain the results tabulated in Table 2. The resulting histogram is shown in Figure 21(b).

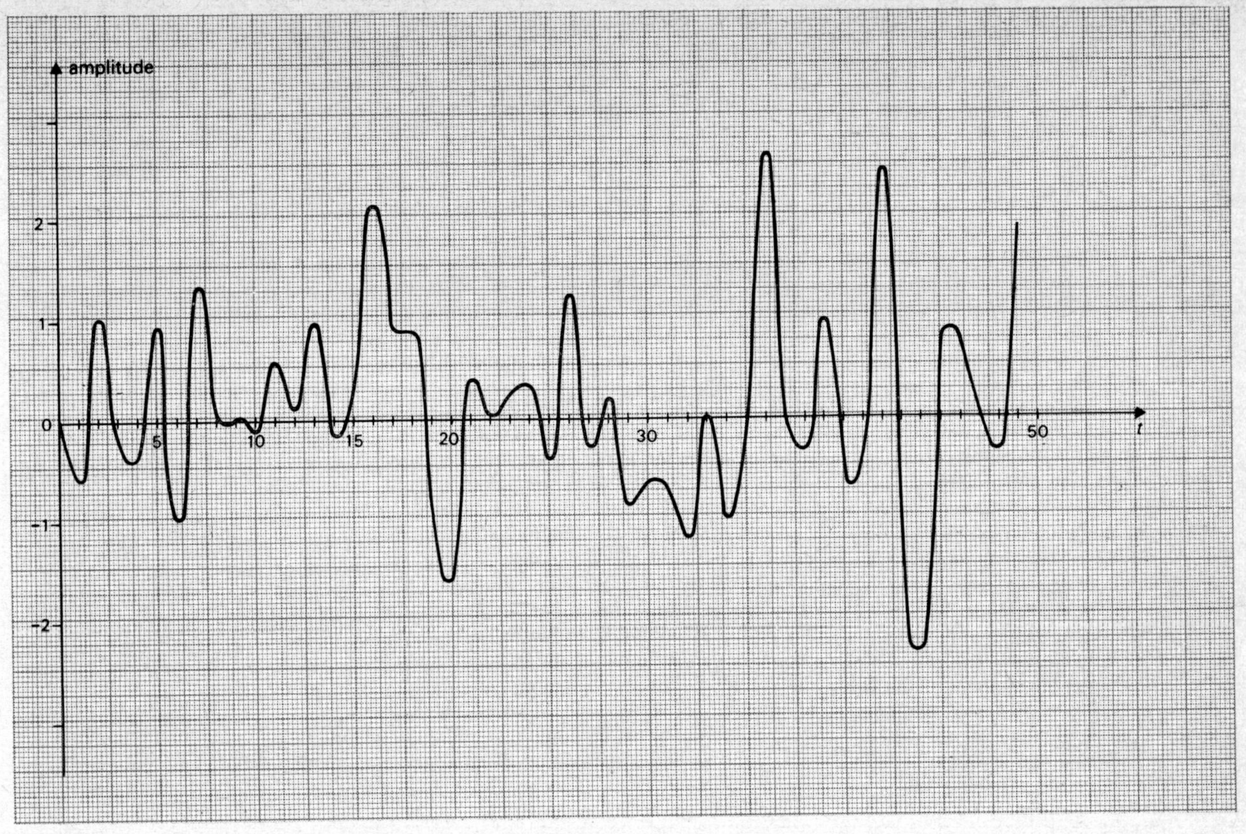

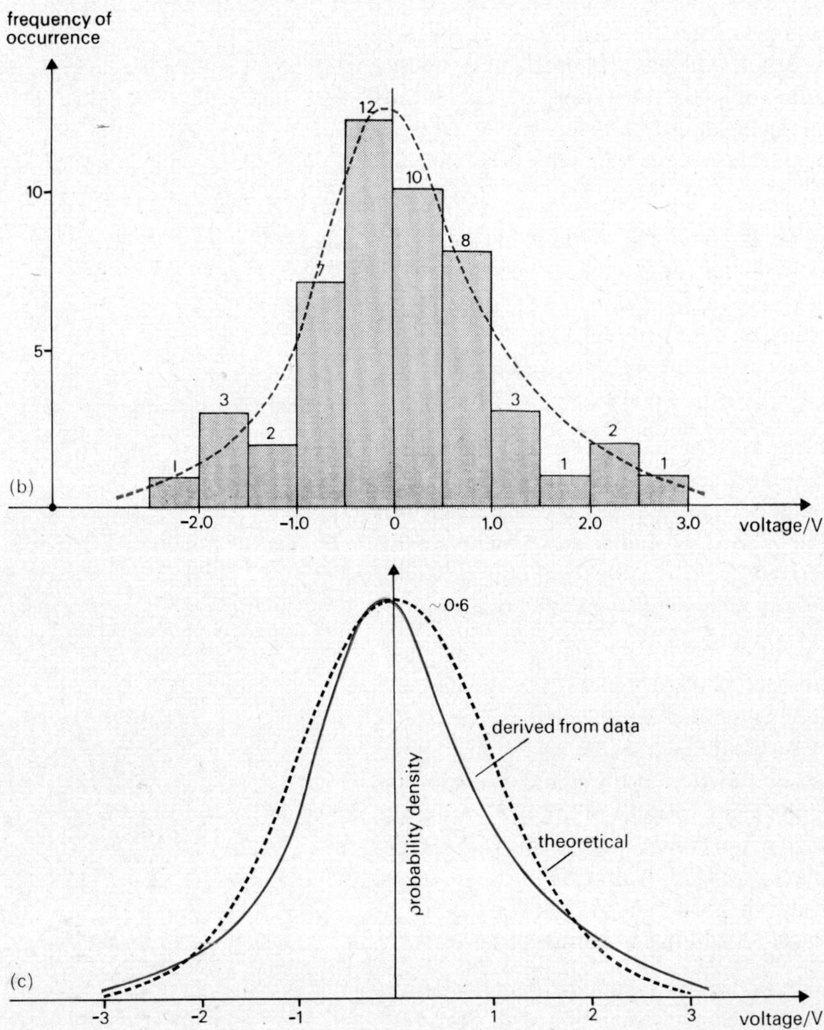

Figure 21(a) A segment of Gaussian noise. (b) The histogram of voltage for the Gaussian random waveform of (a). (c) The derived probability density function and the theoretical Gaussian, or normal, distribution

Table 2 Fifty sample values of Gaussian noise whose mean is zero and whose r.m.s. value is 1.0

−0.59	1.02	−0.25	−0.34	0.95
−1.00	1.36	0.03	0.04	−0.13
0.65	0.11	0.99	−0.17	0.39
2.15	0.91	0.89	−1.43	−1.69
0.42	0.05	0.26	0.33	−0.42
1.24	−0.30	0.21	−0.89	−0.63
−0.71	−1.23	0.03	−1.00	−0.06
2.61	−0.08	−0.33	0.99	−0.61
−0.25	2.47	−1.97	−2.26	0.77
0.79	−0.07	−0.32	−0.66	1.90

The final step is to estimate the probability density function of the signal by drawing a smooth curve through the distribution, as shown in Figure 21(b).

What, then, is the essential difference between a histogram and a probability density function?

We can think of a probability density function as a theoretical or ideal limiting form of a histogram where the number of samples taken is very large. Thus, applied to a signal, histograms are the results of measurement on waveforms, or segments of a signal, whereas the probability density function is inferred from these results – or alternatively derived theoretically – and refers to the whole signal. There is only one probability density function for a signal, though measurements on segments of it can give rise to many, similar, histograms.

Now, the dashed curve in Figure 21(b) is the best estimate of the probability density function based on the results of the measurements. But, since this was a very limited sample, we might distrust its form somewhat and infer from the results that it is indeed Gaussian noise. If this is the case, we can draw the theoretical probability density function, the so-called *Gaussian distribution* shown in Figure 21(c).

Gaussian distribution

You see, we can never actually *measure* a probability density function of a signal. If we were to quantify the whole signal, there would no longer be any uncertainty as to what it contains: it would be fully described by a (large) histogram. If, on the other hand, we quantify segments of the whole signal (i.e. waveforms derived from it), we can only *infer* the properties of the whole signal, so any statement about the signal can be no more than probabilistic.

In practical instrumentation situations it is often unnecessary to make this distinction. There may be no significant difference between one set of sampled data derived from a signal and another set, so all the histograms are the same as the probability density function of the signal. But suppose, for example, a meter is indicating the r.m.s. voltage of a noise source and its reading is fluctuating .These fluctuations reveal that in this case there *is* a significant difference between one set of data within the meter response time and the next. The best estimate of probability in such a case is the mean value of the meter readings. But it still has a margin of uncertainty.

The Gaussian distribution is such an important type of probability density function that we shall spend quite some time discussing it in Section 7. Before then, however, I shall summarize the results of Section 6.

6.6 Summary

1 The mean value v_m of n sample values $v_1, v_2, v_3, \ldots, v_n$ of a signal is

$$v_m = \frac{v_1 + v_2 + v_3 + \ldots + v_n}{n}.$$

2 The standard deviation σ of n sample values $v_1, v_2, v_3, \ldots, v_n$ of a signal is

$$\sigma = \sqrt{\frac{(v_1 - v_m)^2 + (v_2 - v_m)^2 + \ldots + (v_n - v_m)^2}{n}}. \quad (2)$$

3 If the mean value of a signal is zero, the r.m.s. value of the signal is equal to the standard deviation of the sample values of the signal.

4 A histogram of sample values of a random waveform is a plot of the frequency of occurrence of sample values within each of a number of specified ranges of values, plotted against the midvalues of the ranges.

5 The probability density function is a statement, sometimes represented as a graph, of the likely frequency distribution of the whole signal. It may be inferred from measurements or derived theoretically.

6 The area under the graph of the probability density function is unity.

7 The sample values of Gaussian noise can, by definition, be described by a Gaussian probability density function.

The Gaussian, or normal, distribution

The terms 'Gaussian distribution' and 'normal distribution' are synonymous. The reason why the Gaussian distribution is important is that it turns out to be a good probability density function for describing many naturally occurring or artificially produced phenomena. It describes a *normal* situation. For instance, the heights of individuals in a population are distributed in this way around the mean. Weights of manufactured articles are often distributed in this way about their mean weight.

There is, in fact, a mathematical theorem which states that this should be so. The theorem is called the central limit theorem and, briefly, it states that the *sum* of independent but like events or quantities occurring randomly always tend towards a 'normal' distribution. This is true whether or not the summed events are themselves 'normally' distributed. We shall be returning to this definition later, but first let us take a simple illustrative problem.

Suppose we have the task of combining 10 Ω resistors to make 20 Ω, 30 Ω, 40 Ω, 50 Ω, . . . resistors. Obviously we can connect an appropriate number of 10 Ω resistors in series for this purpose. Suppose, however, that the box of so-called 10 Ω resistors contains only 9 Ω and 11 Ω ones. They are all marked as 10 Ω and there is no way of telling a 9 Ω from an 11 Ω resistor just by looking at them, but we know they are in equal numbers, so the chance of choosing either is the same.

First consider making a 20 Ω resistor by putting in series two nominally 10 Ω ones chosen from the box.

> What will be the distribution of resistance values of 20 Ω resistors made up by taking two such resistors and wiring them in series?

We can quickly obtain the result we want by tabulating all the possible combinations, as in Table 3.

Table 3

Nominally 10 Ω resistors		Total resistance/Ω
1st resistor/Ω	2nd resistor/Ω	
9	9	18
9	11	20
11	9	20
11	11	22

So, if resistors are drawn out of the box at random and wired together, there will, on average, be twice as many 20 Ω resistors as there are 18 Ω resistors or 22 Ω resistors.

Now consider making up 30 Ω resistors instead of 20 Ω ones. We would connect three 10 Ω resistors in series. The spread of values would now become that shown in Table 4.

Table 4

Nominally 10 Ω resistors			Total
1st resistor/Ω	2nd resistor/Ω	3rd resistor/Ω	resistance/Ω
9	9	9	27
9	9	11	29
9	11	9	29
11	9	9	29
9	11	11	31
11	9	11	31
11	11	9	31
11	11	11	33

This time four possible values emerge, 27 Ω, 29 Ω, 31 Ω and 33 Ω; and they are likely to turn up in the ratios of 1:3:3:1.

Now, we can continue this investigation further, finding the distribution of resistance values when we make up 40 Ω, 50 Ω, 60 Ω resistors, and so on. For each case the minimum value is obtained from a series of 9 Ω resistors and the maximum from a series of 11 Ω resistors, so the final value is never more than ±10 per cent in error. The probabilities of made-up resistances with less than ±10 per cent errors increase steadily as the number of nominally 10 Ω resistors added together increases.

The distributions can be tabulated in Table 5. This array of numbers is often referred to as Pascal's triangle.

Table 5 Pascal's triangle

Nominal resistance	Relative probabilities	Total
10 Ω	1 1	2
20 Ω	1 2 1	4
30 Ω	1 3 3 1	8
40 Ω	1 4 6 4 1	16
50 Ω	1 5 10 10 5 1	32
60 Ω	1 6 15 20 15 6 1	64
70 Ω	1 7 21 35 35 21 7 1	128
80 Ω	1 8 28 56 70 56 28 8 1	256

Each number in the pyramid is obtained by adding the two numbers nearest to and above it. The meaning of a row in the pyramid can be illustrated using any one row. We have already looked at the 20 Ω and 30 Ω cases. Now let us look at the 50 Ω case

The two extreme values are 45 Ω and 55 Ω. Each other possible value must differ by 2 Ω from its nearest neighbour. So the range of values must be

45, 47, 49, 51, 53 and 55 Ω.

The pyramid tells us that for every 45 Ω resistor there will be:

five 47 Ω resistors,
ten 49 Ω resistors,
ten 51 Ω resistors,
five 53 Ω resistors,
one 55 Ω resistor.

The numbers in any row of Pascal's triangle are the coefficients of what is called the binomial expansion.* So the frequency distribution represented by the rows in Table 5 are usually called *binomial distributions*.

binomial distribution

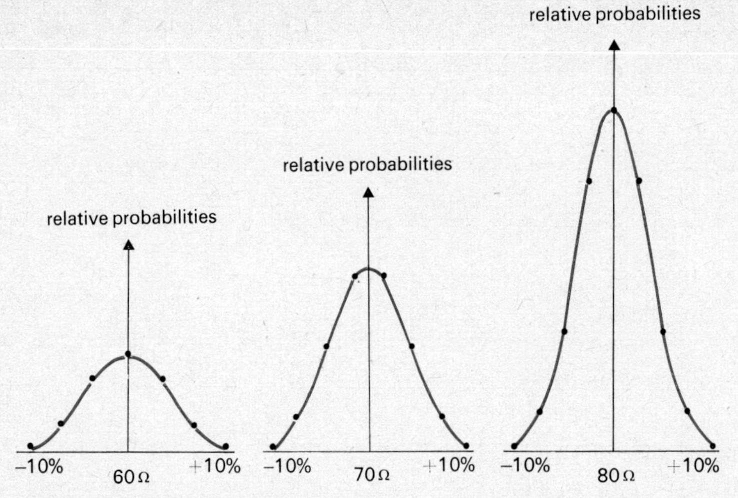

Figure 22 The numbers in the last three rows of Table 5 plotted as a function of percentage deviation from the mean values of 60 Ω, 70 Ω *and* 80 Ω *resistors*

The last three rows of Table 5 are plotted in Figure 22. Each plot or distribution has one sample on each 10 per cent limit. Within these limits a smooth curve can be drawn through the points, as shown in red. The shape of this curve approximates, already quite closely, to the normal, or Gaussian, distribution. If we were to add together more and more resistors – making up larger resistor values – we would obtain a set of points lying closer and closer to the Gaussian form.

The main point to notice is that this Gaussian distribution has emerged from the addition of a random selection of resistors with only two possible values. (Such resistance values form what is called a 'two-valued' variable.) Resistors of only 9 Ω and 11 Ω were considered, yet there emerges a normal, or Gaussian, distribution. In fact, any distribution of so-called 10 Ω values in the box, having the same mean, will lead to a similar Gaussian distribution. This illustrates why the Gaussian distribution is so commonly encountered.

Let us take another simple distribution. Suppose that our original box of so-called 10 Ω resistors contained a uniform distribution of three values, that is, it contained 9 Ω, 10 Ω and 11 Ω resistors in equal proportions. What distributions of values of series resistors would result? This time the table for a 20 Ω resistor appears as in Table 6.

Table 6

Nominally 10 Ω resistors		Total resistance/Ω
1st resistor/Ω	2nd resistor/Ω	
9	9	18
9	10	19
9	11	20
10	9	19
10	10	20
10	11	21
11	9	20
11	10	21
11	11	22

**If you take the coefficients of powers of x (remembering that $1 = x^0$) in the following products, you obtain the numbers in the rows of Table 5 (these are called binomial expansions):* $1+x$, $(1+x)(1+x)$, $(1+x)(1+x)(1+x)$, *etc. The last row in Table 5 contains the coefficients of* $(1+x)^8$.

We now have five possible values, 18, 19, 20, 21 and 22 Ω and they will, on the average, appear in the ratios

1:2:3:2:1.

To make a nominally 30 Ω resistor we have to add one more nominally 10 Ω resistor, which has three equally probable possible values. Now, an exactly 30 Ω resistor can be made from any 19 Ω, 20 Ω or 21 Ω resistor. Since these occur with likelihoods of 2, 3 and 2 as compared with an 18 Ω resistor, it follows that an exactly 30 Ω resistor will occur seven times (2+3+2) more often than a 27 Ω resistor.

We can build up a new pyramid in which each number is obtained by adding together the three numbers immediately above it, as shown in Table 7. This distribution is compared with the previous one in Figure 23.

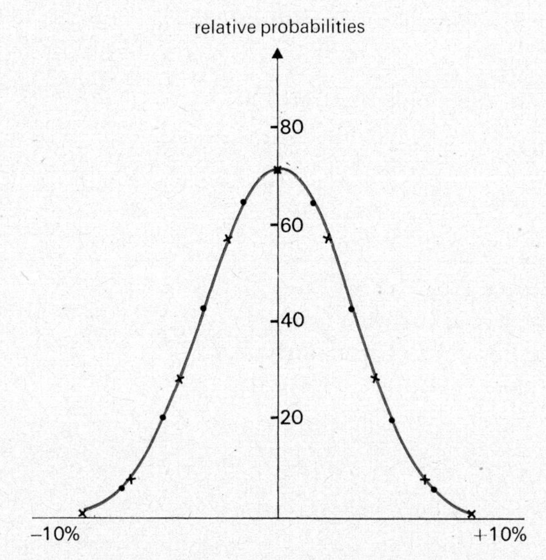

Figure 23 The distribution resulting from combinations of two sets of resistors: crosses represent the 80 Ω resistor from Table 5; dots represent the 50 Ω resistor from Table 7, scaled by a factor of 7:5. The theoretical Gaussian density function is shown in red

Table 7

Nominal resistance	Relative probabilities											Total
10 Ω					1	1	1					
20 Ω				1	2	3	2	1				9
30 Ω			1	3	6	7	6	3	1			27
40 Ω		1	4	10	16	19	16	10	4	1		81
50 Ω	1	5	15	30	45	51	45	30	15	5	1	243

(As a matter of fact, we would still obtain a Gaussian distribution if the mean values of the initial boxes of resistors were not the same, but then, of course, the mean of the final distribution would not be a multiple of 10 Ω.)

This illustration using series resistors is particularly easy to describe. We would obtain the same result if we took resistors in parallel and made 5 Ω, $3\frac{1}{3}$ Ω, 2.5 Ω, etc. resistors. We find the same distribution of many other composite items.

Consider, for instance, thermal noise, the electrical noise which is caused by the random motion of electrons in a resistor. The value of the noise voltage at any instant is the sum of the effects of the random motion of a very large number of electrons. It is not surprising, therefore, to find that it has a Gaussian distribution. Shot noise also has a Gaussian distribution because the net voltage is the sum of individual contributions of a large number of electrons.

So, the Gaussian distribution is a very pervasive distribution. It describes the distribution of the values of many types of electrical noise, as well as the variation of weight, size, length, etc. of many kinds of article we come across.

As I mentioned previously, this general *convergence* upon the Gaussian distribution is called the *central limit theorem.* A more precise statement of the theorem is that

The probability distribution of the sum of n random variables, each of which has its own probability density function, will approach the Gaussian density function as n becomes large.

This is a more general statement than the resistor example illustrated.

In what way is it more general?

In the resistor example each further 10 Ω resistor added came from the same set of resistors. The theorem states that we would still get a Gaussian distribution of resistance values if we took the first resistor from one bin (e.g. 9 Ω, 11 Ω), the second resistor from another (9 Ω, 10 Ω, 11 Ω), the third from another, and so on.

What, then, can we say about the properties of this very pervasive probability density function?

The Gaussian probability density function is shown in Figure 24. It can be fully specified by two parameters:

1 the mean x_m;

2 the standard deviation σ.

We have come across both of these terms before, in Section 6 of this unit. Check back if you are unsure of their meanings.

Since this curve represents a *probability density* and, as I have said before, the sum of all possible probabilities must add up to certainty – which is a probability of 1 – the area under the curve must be 1. In Figure 24, with the x-axis marked out in units of σ (the standard deviation), the probability density measured along the y-axis reaches a maximum of

$$\frac{1}{\sqrt{(2\pi)}}=0.399.$$

One standard deviation away from the mean turns out to correspond to a probability density of about 0.6 of the maximum.

For reference only, I include the equation for the Gaussian distribution with zero mean. It is

$$y=\frac{1}{\sigma\sqrt{(2\pi)}}\exp\left[-\frac{x^2}{2\sigma^2}\right].$$

The probabilities of occurrence of values between $\pm\sigma$, $\pm2\sigma$ and $\pm3\sigma$ are shown in Figure 24. About 68 per cent of the time the values of a continuous Gaussian random variable such as Gaussian noise will be within $\pm\sigma$ of its mean value. Similarly, about 95 per cent of the time it will be between $\pm2\sigma$ and more than 99 per cent of the time between $\pm3\sigma$.

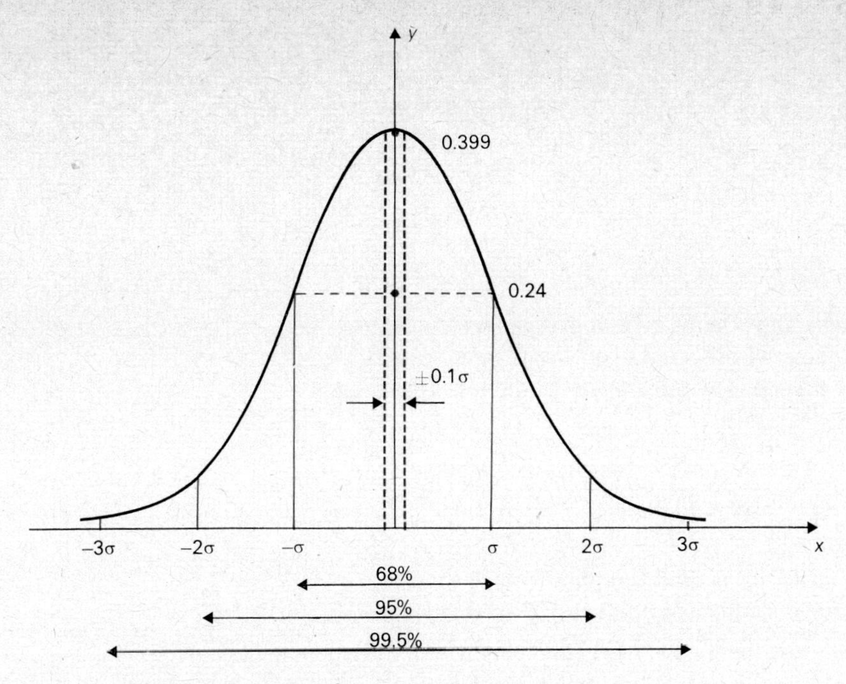

Figure 24 The Gaussian density function, or Gaussian distribution

What is the probability of the value of such a variable lying within $\pm 0.1\ \sigma$ of the mean value?

We must find the area of the strip marked on Figure 24. This is about $0.399 \times 0.2 = 0.08$.

SAQ 6

SAQ 6

Sketch for yourself the probability density function of Gaussian noise whose r.m.s. voltage is 0.2 V. Mark out the axes. In particular, mark the peak value of the probability density. What proportion of the time can you expect the signal value to exceed +0.4 V.

The $\pm 2\sigma$ and $\pm 3\sigma$ ranges are the basis of what are sometimes called the 95 and 99 per cent 'confidence limits'. Since a Gaussian random variable might, in principle, take any value – even though some values are extremely rare – it is not possible to say absolutely that any value will or will not occur. So one says, for example, of a batch of resistors that their resistances are between ± 5 per cent of their nominal value within the '95 per cent confidence limit'. This means that, if the resistor values have a Gaussian distribution, the standard deviation is about 2.5 per cent, since the '95 per cent confidence limit' corresponds to $\pm 2\sigma$.

What are the limits of variation of these resistors within the 99 per cent confidence limit?

The 99 per cent confidence limit corresponds to about $\pm 3\sigma$, so the limits on resistance tolerance are ± 7.5 per cent within the 99 per cent confidence limit.

Thus the probability density function is a convenient way of describing the likelihood of occurrence of any range of values of a particular variable.

There is one important difference between random variables like a man's height or a signal voltage and random variables like the resistance of a set of resistors made up from connecting particular values in series, as described earlier.

Can you see what it is?

One is a *continuous random variable*, the other is a *discrete random variable*. A signal voltage can have *any* value within prescribed limits, and so we speak of probabilities of values occurring within *ranges* of values. These values form a *continuous* random variable.

The resistors we considered earlier had a range of *particular* resistances; no intermediate values were possible. So these values form a *discrete* random variable – only discrete or distinct values are available.

The phrase 'probability density function' is used to apply to each kind of variable because, of course, if there is a sufficient number of discrete values of a variable within the range it becomes indistinguishable from a continuous random variable.

The main difference between the density functions is that one can be represented by a continuous graph, like a Gaussian density function, while for the discrete variable the density function is a set of numbers (their probabilities) associated with the discrete values (e.g. the numbers in Pascal's triangle).

It is the *envelope* of these values which can be represented by a continuous graph. Evidently, too, for a discrete random variable, you do not measure the area under the curve to obtain probabilities, the probabilities are already tabulated.

SAQ 7

Explain the difference between a histogram of values of a waveform and the probability density function of a signal.

SAQ 7

7.1 The average standard deviation

Finally, we must consider a property of the Gaussian distribution which is most important from the point of view of our case studies. That is the way in which the standard deviation of the *sum* of two or more similar Gaussian distributions is related to the standard deviation of each. We shall find that, by *averaging* the random values of two or more noise waveforms, the effective or average standard deviation of the noise decreases. On the other hand, the average of a *repeated signal* does not decrease. These two factors make it possible to diminish the effects of noise upon a wanted signal.

Again, the easiest way of understanding this is to consider the simple example of adding resistors together. Table 5 shows the distribution of resistance values of resistors whose nominal values are 10, 20, 30, . . . , 80 Ω. Now, for the purposes of the present problem we shall think of the 80 Ω resistors as being formed by adding together two 40 Ω resistors, each of which has the probability density function given in Table 5, namely:

Resistance/Ω	36	38	40	42	44
Frequency of occurrence	1	4	6	4	1

(*Note.* This is a *discrete* probability density function.)

SAQ 8

SAQ 8

What is the standard deviation of the distribution of nominally 40 Ω resistors tabulated above?

You should have obtained the answer 2 Ω for this standard deviation. Now, the probability density function of resistance values of the 80 Ω resistor is given in Table 5, namely:

Resistance/Ω	72	74	76	78	80	82	84	86	88
Frequency of occurrence	1	8	28	56	70	56	28	8	1

The standard deviation of this distribution can be similarly worked out and comes to

$$\sigma_{80}=2\sqrt{2}\ \Omega,$$

where σ_{80} stands for the standard deviation of the set of 80 Ω resistors.

Thus the standard deviation has increased. In fact, by adding together two identical distributions, we have increased the standard deviation by $\sqrt{2}$.

By what factor do you suppose the standard deviation will increase when three of those distributions are added together?

It is a simple matter, though a tedious one, to calculate that, for the resulting 120 Ω resistors,

$$\sigma_{120}=2\sqrt{3}\ \Omega.$$

Indeed, the standard deviation *increases as the square root of the number of distributions added together.*

If 100 distribution were added together, by how much would the standard deviation increase?

The standard deviation would increase by $\sqrt{100}$, that is, by 10.

Now, although the standard deviation is increasing as the number of samples is increased, so also is the total resistance. Indeed, as already explained, the *percentage* spread of resistor values never exceeds ± 10 per cent of the nominal value, even though the *total spread* in resistance increases in proportion to the resistance. Thus the spread of the 80 Ω resistors is ± 8 Ω, the spread of 120 Ω resistors is ± 12 Ω.

By how much does the standard deviation change, *expressed as a percentage of the nominal value*, as the nominal value increases? This is called the *average standard deviation* σ_{average}.

The nominal resistance increases in proportion to the number n of distributions added together.

The standard deviation increases in proportion to $\sqrt{n}$.

So σ_{average} is proportional to $(\sqrt{n})/n=1/\sqrt{n}$. That is, the *average standard deviation decreases by* $1/\sqrt{n}$ *as the number of distributions added together increases.*

The result has been derived for the case of binomially distributed variables. It also applies to Gaussian random sources.

Now, this idea of an 'average standard deviation' is rather a difficult one to be clear about, so let me repeat the points one by one. This time, though, I shall consider a voltage source with a random noise added to it (rather than resistors with varying tolerance).

1 First we take samples of this random source – say 100 samples, to be specific. We find that these samples have a mean value of $\bar{v}$ and Gaussian distribution with a standard deviation of σ_{source}.

Suppose, for example, that

$$\bar{v}_{source} = 10 \text{ V},$$
$$\sigma_{source} = 1 \text{ V}.$$

2 We next take a second set of 100 samples and again find they have a similar Gaussian distribution.

Thus, again, $\bar{v}_{source} = 10$ V,
$\sigma_{source} = 1$ V.

3 We now add these two sets of samples togethei by repeatedly adding together one sample from each distribution. We shall finish up with 100 additions of pairs of samples.

We now find that the mean of the sum has doubled while the standard deviation has only increased by $\sqrt{2}$. That is,

$$\bar{v}_{sum} \approx 20 \text{ V},$$
$$\sigma_{sum} = \sqrt{2} \text{ V}.$$

4 We can now work out the *average* values of both the mean and the standard deviation. For example, the *average* standard deviation is this

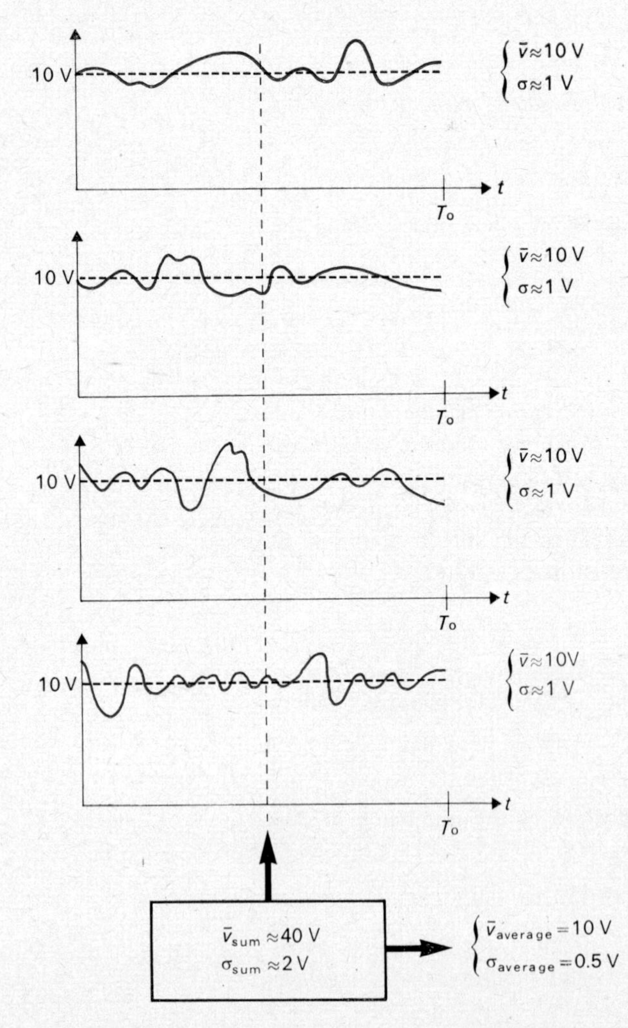

Figure 25 The principle of signal averaging. Each of four waveforms has the same mean and standard deviation. Adding and averaging the values corresponding to a particular part of the waveform results in the same mean but reduced standard deviation

overall standard deviation divided by the number of distributions added together to produce it. Thus, in this example,

$$\bar{v}_{\text{average}}=\frac{20\text{ V}}{2}=10\text{ V},$$

$$\sigma_{\text{average}}=\frac{\sigma_{\text{sum}}}{2}=\frac{\sqrt{2}\text{ V}}{2}=\frac{1}{\sqrt{2}}\text{ V}.$$

5 If the process of taking sets of samples and adding them together is repeated, so that n sets are added together, then, in this example,

$$\bar{v}_{\text{average}}=\frac{n\times 10\text{ V}}{n}=10\text{ V},$$

$$\sigma_{\text{average}}=\frac{\sqrt{n}}{n}\text{ V}=\frac{1}{\sqrt{n}}\text{ V}.$$

So the average standard deviation decreases, though the mean value of the signal does not. Thus this technique makes possible the reduction of the noise component of a signal (measured by σ_{average}) relative to the constant component (measured by $\bar{v}_{\text{average}}$).

This whole process, for the case of $n=4$, is illustrated diagrammatically in Figure 25.

Notice that the quantity I have called $\bar{v}_{\text{average}}$ *is an average of an average!* The average value of one set of samples – 100 in the example given – is $\bar{v}$ and $\bar{v}_{\text{average}}$ is the average of a number of values of $\bar{v}$.

Now let us consider another example.

Measurements are made on a signal in order to quantify it. It is found that all sample values lie in the range from 6.5 V to 13.5 V, so the whole range is divided into seven subranges, each of 1 V width. Thus the subranges centre upon the values 7 V, 8 V, 9 V, 10 V, 11 V, 12 V, 13 V, as shown in Figure 26(a).

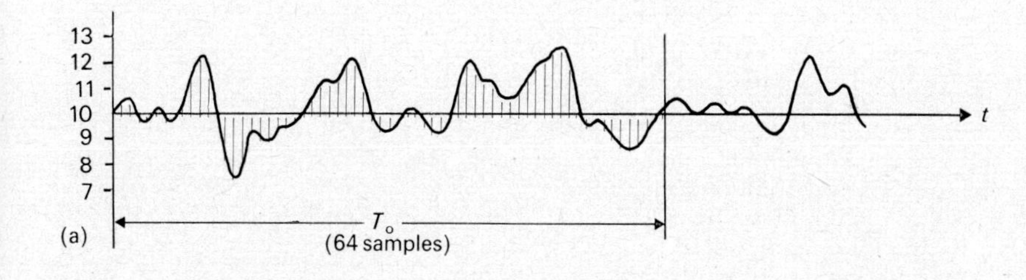

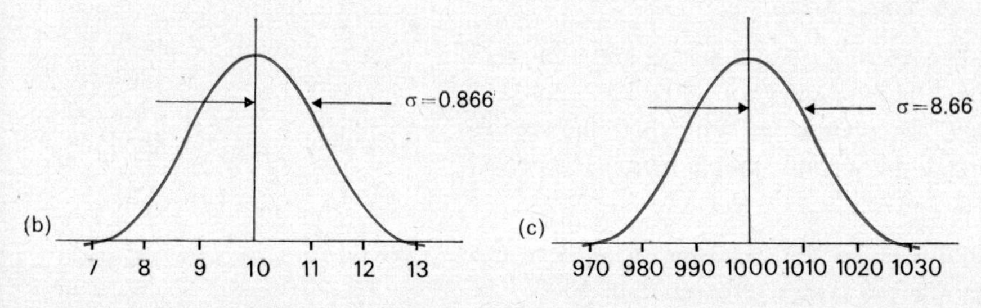

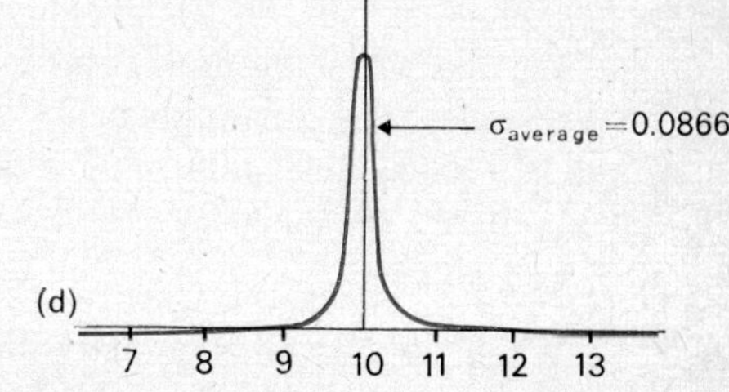

Figure 26 Signal average: (a) A waveform of duration T_o, drawn from a signal; (b) the probability distribution for any 64 *samples; also for* 100 *times* 64 *samples, i.e.* 6400 *samples; (c) the probability density functions for* 64 *summations of the samples taken from* 100 *observations; (d) the average value and average deviation obtained by dividing (c) by* 100

Sixty-four samples are taken in the first observation period of T_o. They are distributed as follows:

Measured value/V	7	8	9	10	11	12	13
Frequency of occurrence	1	6	15	20	15	6	1

I have chosen binomial numbers to keep the arithmetic simple. The mean of this distribution is

$$\bar{v}=10 \text{ V}.$$

The standard deviation is

$$\sigma=\sqrt{\frac{3^2+6\times 2^2+15+15+6\times 2^2+3^2}{64}}$$

$$=\sqrt{\frac{96}{64}}$$

$$=0.866 \text{ V}.$$

This is illustrated in Figure 26(b).

A further sixty-four samples are taken, and although they do not appear in the same order, the distribution is identical to the previous one. So again,

$$\bar{v}=10 \text{ V},$$
$$\sigma=0.866 \text{ V}.$$

In all, the signal is sampled during 100 observation times. Each time the same distribution is obtained, so in each period the mean is 10 V and the standard deviation is 0.866 V.

Evidently, the average of the means is

$$\bar{v}_{\text{average}}=10 \text{ V} \quad \text{again.}$$

Also, the average of the standard deviation is

$$\frac{0.866\times 100}{100}=0.866 \text{ V} \quad \text{again.}$$

Now consider all 6400 samples as a whole, with probability distribution as follows:

Measured value/V	7	8	9	10	11	12	13
Frequency of occurrence	100	600	1500	2000	1500	600	100

I hope you can see without further calculation that once again

$$\bar{v}=10 \text{ V}$$

and $\quad \sigma=0.866$ V.

Now we come to the process called signal averaging.

We take the first sample value from each observation period and add them together. The sum might be 7+11+10+9+10+8+9+10+12+13+11+ . . . (100 samples total) . . . +10 V. On average we can expect the sum to be 10 V×100=1000 V. Then we take the second sample from each period and so on.

When we do this sum sixty-four times we find a range of values. The smallest cannot be less than 700 nor the largest greater than 1300. The standard deviation of these sixty-four values turns out to be 8.66 V (i.e. exactly $\sigma\sqrt{100}$) in this example, as shown in Figure 26(c).

Thus, finally, $\bar{v}_{\text{average}}=1000/100=10$ V,
$\sigma_{\text{average}}=8.66/100=0.0866$ V.

The final distribution is shown in Figure 26(d); a much narrower distribution than in Figure 26(b). Thus the noise component has been decreased in relation to the mean value of 10 V.

This example is, of course, artificial because, for simplicity, I have stated that each set of sixty-four samples has *precisely* the same probability distribution. With a truly random signal each set of sixty-four samples would have a somewhat different mean and standard deviation (perhaps 9, 8.5, 9.5, 10.5, 9, 11, etc. for the mean; 0.8, 0.9, 0.75, 0.95, etc. for σ).

What would be the effect on the calculations of taking, say, 6400 samples within each observation time, instead of sixty-four?

The spread in values of the means and the standard deviations for each observation period would be less. Each mean would be closer to the 'true' mean of 10 V, and each calculated standard deviation would be closer to 0.866 V.

The effect of averaging at each sample instant would not be affected.

7.2 Summary

1 Gaussian distributions are those histograms which are well described by the Gaussian probability density function.

2 Sixty-eight per cent of the Gaussian distribution is within $\pm\sigma$. Ninety-five per cent is within $\pm 2\sigma$. Over 99 per cent is within $\pm 3\sigma$.

3 When n separate Gaussian distributions are added the overall standard deviation increases as $\sqrt{n}$.

4 Thus the 'average standard deviation' – which is the sum divided by n – decreases as $1/\sqrt{n}$.

5 This process forms the basis of the technique of signal averaging for improving signal-to-noise ratio.

Noise and signal averaging

Now let us consider how this last property of noise – that its standard deviation decreases as you add portions of it together and take their average – can be used in the detection of signals.

phase-sensitive detection
synchronous detection averaging

The method takes various forms, and may be called *phase-sensitive detection* or *synchronous detection* or sometimes simply *averaging*. The method can be used in connection with case study 1, described briefly in section 4.2, which is concerned with measuring the evoked response of a nervous system.

The basic idea is to make the response we want to study into a periodic signal. The response must be of finite duration, namely, the observation time T_0, but within this period and within the bandwidth of the system, the response is a random waveform. So also is the interfering noise, but we have to separate the two. The trick is to generate a repeated version of the wanted response, thereby producing a periodic signal (see Figure 13). Thus the overall signal is a repeated version of the wanted response, mixed in with random noise. By *adding together* a series of such responses and taking the average, the wanted waveform is maintained while the noise tends to average out.

The periodic signal is generated by applying a pulse to the network to be studied, in this case part of the nervous system, and the 'evoked' response is what would be observed somewhere else in the network if there were no noise. By repeatedly applying the pulse to the network, repeated versions of the wanted response are obtained, mixed in with noise. As indicated in Figure 27, the noise may be so great as almost completely to mask the wanted evoked response.

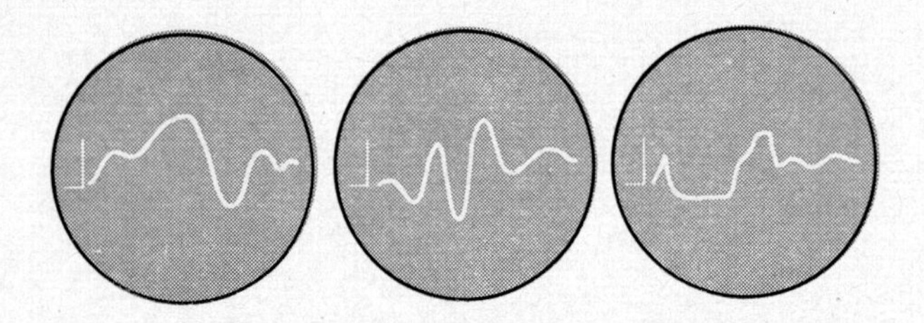

Figure 27 Three examples of the direct display of an evoked response from the nervous system. Due to the presence of noise in the system, no discernible characteristic response can be seen

This impulse is applied both to the network whose response we want to study – the hand, arm and head in Figure 8 – and to the trigger input of an instrument called an averager, as shown in Figure 28.

The averager comprises both an oscilloscope and a simple computer. The computer part adds the signals together and the oscilloscope displays either what is in the computer store or what is applied directly to its signal input.

The impulse applied to the network 'evokes' the network response at its output (the scalp, in our example). The impulse applied to the averager triggers an oscilloscope sweep, and, if the output of the network is connected to the signal input of the averager, the evoked response of the network will be shown on the oscilloscope screen.

So far the operation is typical of any oscilloscopic display of circuit

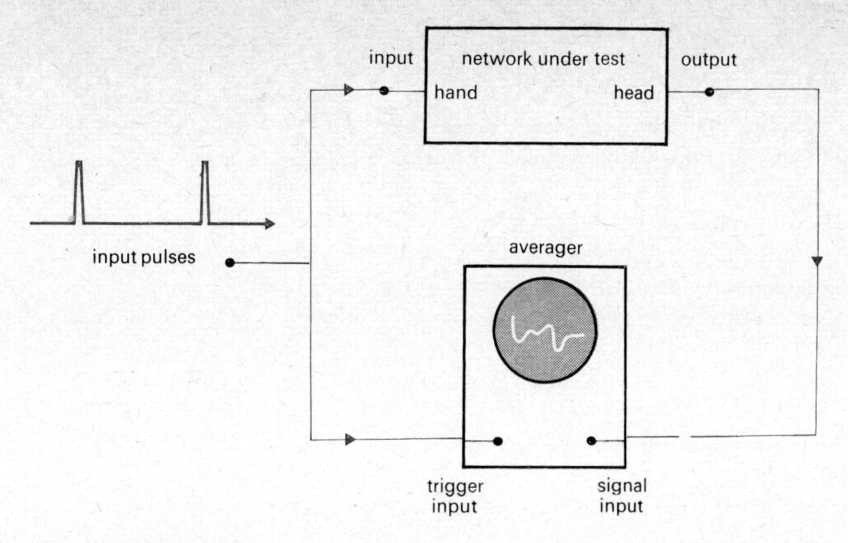

Figure 28 The arrangement of a signal averager. The input pulses both excite the network and trigger the averager

response to a signal. The signal drives the network and starts the oscilloscope trace. The oscilloscope then displays the network response each time the impulse is applied. If the network response is noise free, a clear display of impulse response of the network will be seen, but in this case noise is added, so each response will be different, as already indicated in Figure 27.

The averager does not just take each signal as it comes, it adds each signal it receives to the signals it has previously received, so that in effect each network response is 'laid on top of' the others it has received and it displays the sum of them all, just as indicated in Figure 25.

It also divides the sum it obtained by n, the number of responses it has received, so that the response it displays is the average of n network responses.

The way it does this is to take a number of samples of each response, say sixty-four samples. Each sample is associated with an 'accumulator', into which the successive samples of each instant are added. Thus the summing process indicated in Figure 25 occurs for each of the sixty-four sample instants. After, say, fifty responses every one of the sixty-four accumulators will then contain the sum of fifty samples of the instantaneous amplitude of the network response. If the content of each accumulator is then divided by fifty, the value resulting will be the network response at the corresponding instant of the sample together with a random added component whose standard deviation has decreased by a factor of $\sqrt{50}$ compared with the original signal.

This applies to each of the sixty-four sample points of the sampled network response. The net result is a display of the evoked response with a reduction in the overall standard deviation, or r.m.s. voltage of the noise, by $\sqrt{50}$. After n responses the noise voltage is in effect reduced by $\sqrt{n}$.

Thus, as you can see on the television programme associated with this unit, as the number of impulses applied to the network increases, the evoked response becomes more and more clearly recognizable. It seems to emerge out of the noise, or, more precisely, the noise seems to fade away. Each response strengthens the periodic signal and weakens the random component.

There is one important assumption implicit in this description of the display of the evoked response. It is that the evoked response is indeed an unchanging response. It is assumed that the response has a consistent variation with time, and that it occurs at a consistent and repeatable delay following the initial response.

If this were not true, what would the system display?

The averager would emphasize whatever *was* consistent and repeatable about the response. Any randomness in the 'actual evoked response' would be regarded as noise, and would be averaged out. Thus it is best, in this instance, to define the evoked response as that which is consistent and repeatable in the network response, and any jitter as part of the noise. (To extract a consistent evoked response waveform which is random in its delay following the impulse involves more sophisticated methods.)

SAQ 9

SAQ 9

What properties must the noise possess if the averager described above is to operate successfully?

If your answer to SAQ 9 included the statement that the noise must be Gaussian, you should think again. Even if the noise were a binary signal (we shall be discussing binary noise later), that is, it has only two values, ± 1 mV for example, the average value of the sum of a random selection of these two values would almost certainly decrease. Only if every random selection were $+1$ mV would the average remain undiminished at $+1$ mV. If any of them were -1 mV then the average would be less than $+1$ mV. Thus the averager will achieve an improvement in signal-to-noise ratio with any kind of stationary noise. The standard deviation will not, however, necessarily diminish initially in proportion to $\sqrt{n}$.

SAQ 10

SAQ 10

Figure 29(a) shows a photograph of Professor Steven Rose undergoing the evoked-response experiment described in this unit. (The experiment was shown as part of a television programme associated with the Open University course SDT286 *The Biological Bases of Behaviour*.) Figure 29(b) shows the same image but reproduced by and photographed on a 'noisy' television set. Figure 29(c) shows a multiple-exposure photograph of six frames of this image, using the same 'noisy' television set. Explain why Figure 29(c) is clearer than Figure 29(b).

SAQ 11

SAQ 11

A sinusoidal waveform has an amplitude of 2 V and can be triggered by an impulse. It is passed through a wide-band network which adds Gaussian noise of r.m.s. value 0.5 V to the waveform. The output and trigger are applied to a signal averager. After how many applications of this noisy waveform will there be only a 5 per cent chance of the averaged waveform being more than 0.1 V in error?

8.1 Summary

When the responses of a system to a series of impulses are summed:

1 The *evoked* responses add linearly, so that the average evoked response stays constant.

2 The noise component adds non-linearly.

(a) If the noise is Gaussian, the standard deviation of the noise contribution increases according to $\sqrt{n}$, where n is the number of samples. In other words, the contribution per sample of the sum (i.e. the average) is proportional to $1/\sqrt{n}$.

(b) If the noise is not Gaussian, its contribution will increase less rapidly than n, but the rate of increase depends upon the probability density function of the noise.

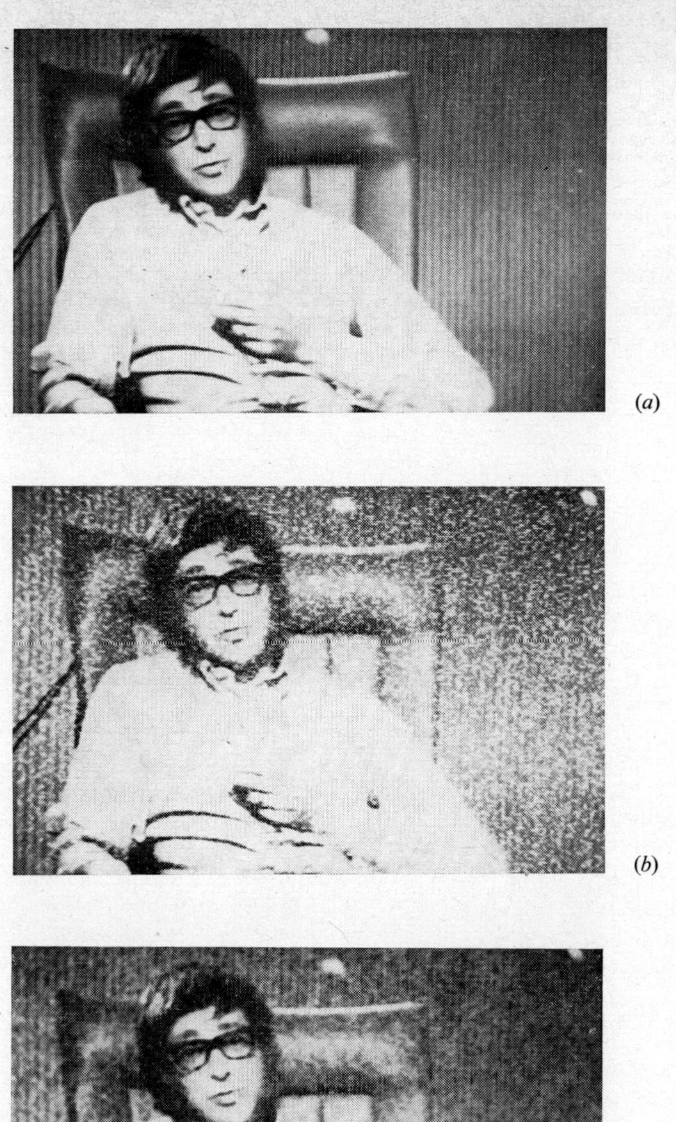

(a)

(b)

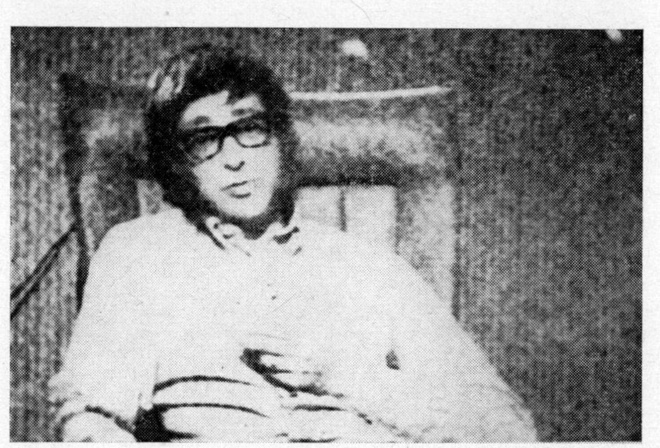

(c)

Figure 29(a) A noise-free picture. (b) The same picture affected by noise. (c) Six exposures of the picture, affected by different segments of noise, averaged together

Self-assessment answers and comments

SAQ 1

The graphs of Figures 4(a–c) plot *attenuation* versus frequency, so:

(a) is a low-pass filter (i.e. zero attenuation at low frequency);

(b) is a band-stop filter;

(c) is a high-pass filter, with a very slow cut-off.

The result of filtering the signal whose spectrum is given in Figure 4(d) is shown in Figure 30.

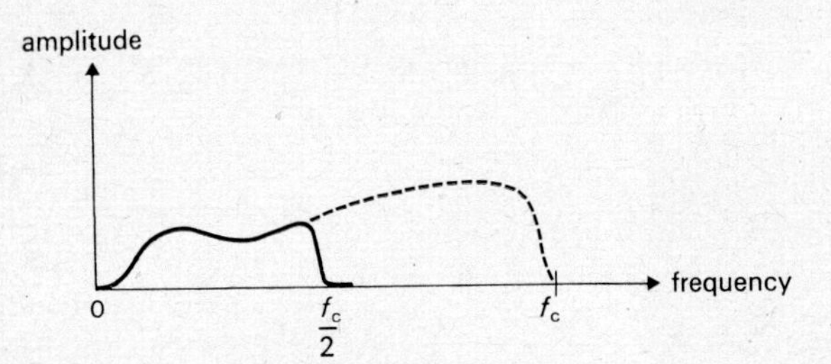

Figure 30 Answer to part of SAQ 1

SAQ 2

The period of the signal generated by repeating the observed waveform is, of course, T_o. Thus the fundamental frequency of the signal is $1/T_o$. The harmonic frequencies must therefore be $2/T_o$, $3/T_o$, . . ., n/T_o.

The maximum frequency within the signal, however, is f_c, so the number of harmonics present in the signal must be given by

$$\frac{n}{T_o} \leqslant f_c,$$

that is, the number of harmonics is given by, at most, f_cT_o.

There are two constants associated with each harmonic: one giving the amplitude, the other defining the phase. Thus the Fourier series is

$$v = A_1 \sin\left[\frac{t}{T_o}+\varphi_1\right] + A_2 \sin\left[\frac{2t}{T_o}+\varphi_2\right] + \ldots + A_n \sin(f_c t+\varphi_n),$$

that is, there are f_cT_o constants giving the amplitudes $A_1, A_2, \ldots, A_n$ and there are f_cT_o phase constants $\varphi_1, \varphi_2, \ldots, \varphi_n$, making a total of $2f_cT_o$ quantities.

SAQ 3

The mean value v_m of a set of n numbers $v_1, v_2, v_3, \ldots, v_n$ is defined as

$$v_m = \frac{v_1+v_2+v_3+\ldots+v_n}{n},$$

so to obtain the mean values you simply add together all the numbers in column A and divide by eighteen, and add together all the numbers in column B and divide by eighteen.

Answers

A. $v_m = \frac{3.74}{18} = 0.207,$

B. $v_m = \frac{1.81}{18} = 0.101.$

These are averages of *waveforms*, that is, they are averages of the segments of signals drawn in Figure 9. What mean value would you *infer* for the continuous sinusoidal *signal* of which this waveform is a part? I think you would (rightly) infer a zero mean for such a symmetrical sinusoid, but our calculation gives a value of 0.207. So, if we want to infer the mean of a continuous repetition of the waveform within the period of a periodic signal, we must average over just one period. In th example given in column A this means averaging over twelv samples. For any twelve consecutive samples the mean is zero.

A similar strategy cannot be adopted for a random waveform because it has no periodicity. The best strategy for inferring the mean of the overall signal from which the waveform wa drawn is to take the mean of as many sample values as are available – eighteen in this case. Even if the mean of a very long sample of the signal were to be zero (to one part in a million, say) it would be very fortuitous if the mean of eighteen samples were zero.

SAQ 4

Here you are asked to *infer* the value of a signal, given only samples of a waveform. Let us assume, then, that the sinusoida waveform is indeed part of a continuous sinusoid, so we must consider only one period, namely twelve sample values.

The r.m.s. value for one period of the sinusoid is

$$\text{R.M.S.} = \sqrt{\frac{\text{sum of squares of 12 samples}}{12}}.$$

The calculation is shown in Table 8.

Table 8

Sample number	Sample value of sine wave	(Sample value)^2
1	0.5	0.25
2	0.87	0.76
3	1.0	1.0
4	0.87	0.76
5	0.5	0.25
6	0	0.0
7	−0.5	0.25
8	−0.87	0.76
9	−1.0	1.0
10	−0.87	0.76
11	−0.5	0.25
12	0	0.0
		6.04

$$\text{R.M.S. value of the sample values of column A} = \sqrt{\frac{6.04}{12}} = 0.71.$$

The r.m.s. value of the noise signal can be found by tabulating *all* its sample values and its (sample values)^2 in the same way. See Table 9.

Table 9

Sample number	Sample values of noise	(Sample value)^2
1	0.5	0.25
2	0.75	0.56
3	0.70	0.49
4	0.78	0.61
5	0.80	0.64
6	0	0.0
7	−0.25	0.06
8	0	0.0
9	−0.15	0.02
10	−0.6	0.36
11	−0.2	0.04

Sample number	Sample values of noise	(Sample value)^2	
12	0.15	0.02	
13	−0.67	0.45	
14	−1.1	1.21	
15	−0.35	0.12	
16	0.2	0.04	
17	0.6	0.36	
18	0.65	0.42	
		5.65	Total

$$\text{R.M.S. value of the noise waveform} = \sqrt{\frac{5.65}{18}} = 0.56.$$

This is the best estimate we can obtain for the r.m.s. value of the whole assumed noise signal.

SAQ 5

To obtain the density function you have first to plot the histogram, that is, the number of sample values within a range against the midvalue of the range. A smooth curve drawn through the points so obtained is an estimate of the probability density function of a sine wave. The numbers of the sample points, shown in Figure 13(a), which fall into the ranges 0–1, 1–2, 2–3, etc. are tabulated in Table 10.

Table 10

Range from	to	Number of sample values	
−5	−4	20	
−4	−3	10	
−3	−2	8	
−2	−1	6	
−1	0	6	
0	1	6	
1	2	6	
2	3	8	
3	4	10	
4	5	20	
		100	Total

The chances of obtaining a sample value in the range from 0 to 1 are 6 in 100 and this is a probability of 0.06. If the probabilities for the other ranges are calculated in the same way and plotted against the midvalue of the ranges, Figure 31 is the result.

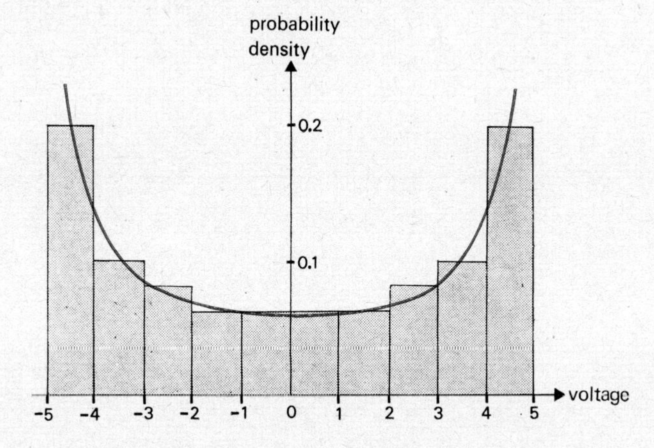

Figure 31 Answer to SAQ 6: the histogram is shown in black and the inferred probability density function is shown in red

SAQ 6

The r.m.s. voltage is the same as the standard deviation, so your curve should be the same as Figure 24 except that each σ should be replaced by 0.2 V.

The area under the whole curve must be unity. The peak probability density in Figure 24 is about 0.4, corresponding to the value of the standard deviation of 1.0. In this case the standard deviation is 0.2 V, the peak value of the curve must be about 2.0 V^{-1} so that the product of the two values remains equal to 1.0.

The signal value 0.4 V is twice the standard deviation, so the *magnitude* of the noise voltage will exceed ± 0.4 V for about 5 per cent of the time. The voltage therefore will exceed $+0.4$ V about 2.5 per cent of the time.

SAQ 7

This is fully explained in Section 6, but it is an important point, since it reflects the difference between experimental results and underlying trends inferred from them.

A histogram representation is a tabulation of the results of measurements. It states the number of times particular values of a discrete variable, or values within particular subranges of a continuous variable, occur.

The probability density function expresses *likelihood* of these discrete values, or subranges of values, occurring. It may be inferred from a histogram or it may be derived from theoretical principles.

SAQ 8

The standard deviation is readily calculated as described in section 6.4.

There are eight resistors whose values differ from the mean by 2 Ω, and two resistors whose values differ by 4 Ω. There are sixteen resistors altogether. Thus

$$\sigma = \sqrt{\frac{2\times 4^2 + 8\times 2^2}{16}}$$

$$= \sqrt{\frac{32+32}{16}}$$

$$= 2.$$

SAQ 9

The noise must have a mean value of zero, be stationary and have no components which are synchronous with the stimulus.

Remember that the averager emphasizes everything that is consistent or repeatable about the response. So, if the mean value of the noise were not zero, this mean value would be emphasized and the full benefits of reducing the r.m.s. value of the noise would not be obtained. Similarly, any component which is synchronous with the stimulus, or a harmonic of it, will be emphasized, since it will occur at the same time in each evoked response.

The noise must also be stationary. If it is not, unpredictable results will be obtained.

SAQ 10

The explanation of the improvement resulting from adding and averaging several noise pictures is similar to that achieved by the signal-averaging process used to extract the evoked response of a nervous system.

By adding together n versions of the noise-free picture and then reducing the overall intensity of the picture by a factor proportional to n, the result is the same as the original noise-free picture.

If the same process is carried out with n noisy pictures, in which the noise each time is a different sample of Gaussian noise, the noise intensity will increase during the adding process according to $\sqrt{n}$. Reducing the overall intensity by a factor of n will therefore cause a reduction of the noise relative to the picture in proportion to $\sqrt{n}$. This is shown in Figure 29(c).

SAQ 11

If the interference due to averaged noise must not exceed 0.1 V within the 95 per cent confidence limit, it follows that the average standard deviation must be 0.05 V (i.e. the 95 per cent confidence limit is twice the standard deviation). The average standard deviation is related to the initial standard deviation by $\sqrt{n}$, the square root of the number of applications of the waveform.

Thus

$$\sigma_{\text{average}} = \frac{\sigma_{\text{source}}}{\sqrt{n}}$$

That is, $0.05 = \dfrac{0.5}{\sqrt{n}}$.

So $n = 100$.

The number of applications of the waveform needed is 100.

Unit 12

Contents

Introduction

In Unit 11, in introducing some of the problems which arise when wanted signals are interfered with by extraneous signals, I defined some terms that I shall be continuing to use throughout the rest of the course.

SAQ 1 (revision)

SAQ 1

What is the difference between a 'random waveform' and a 'deterministic waveform' of limited duration and bandwidth?

I showed that a random waveform of limited bandwidth B and duration T_0 needs $2BT_0$ quantities to describe it completely, but as this might well be an inconveniently large set of numbers, I introduced several parameters which, although not specifying the waveform completely, did describe some of the characteristics of the waveform. Three of these parameters were the mean value, the r.m.s. value and the standard deviation.

SAQ 2 (revision)

SAQ 2

(a) Will mean values determined from different segments of the same noise signal be equal?

(b) What is a stationary random signal?

A measure of the variation of the signal was specified by the probability density function. The Gaussian probability density function was found to be a particularly common form of probability density function.

Finally we saw that 'averaging' can bring about improvement in signal-to-noise ratio in certain situations.

SAQ 3 (revision)

SAQ 3

If a wanted periodic signal and some random Gaussian noise that is superimposed on it have the same r.m.s. value, what will be the ratio

$$\frac{\text{R.M.S. value of the signal}}{\text{R.M.S. value of the noise}}$$

at the output of an averager which averages over 400 periods of the wanted signal?

In this unit I shall continue with the explanation of the parameters which describe the characteristics of random signals. I shall be considering parameters related to the frequency content of the random signals. These parameters are the power density spectrum and the autocorrelation function

These parameters, related to the frequency components of the signal, give further techniques for minimizing the effect of noise upon the signal. I have already mentioned some of the ways in which a knowledge of the spectrum of a signal helps to reduce the effects of noise. As I explained in Unit 11, if a signal's bandwidth is known, the instrumentation system should be designed to pass just those frequencies in the signal. Also, a knowledge of the frequency components of a signal enables us to choose a correct sampling frequency.

I shall begin this unit by extending the idea of being able to represent a periodic signal by a Fourier series to cover signals which are *not* periodic. I shall also explain some of the ways in which the power density spectrum can be measured and I shall be outlining some of the practical problems involved in this measurement

Then I shall give some more examples of matching the system bandwidth to the signal's frequency components so that the effect of interfering noise is diminished

The last of the parameters introduced in Unit 11, the autocorrelation function, is closely related to the spectrum of the signal. I shall spend some time discussing it since it provides one way of obtaining a signal's power density spectrum. Towards the end of this unit I shall introduce some techniques of modulation which allow the transmission of the wanted signal through noisy channels with the noise causing less effect on the wanted signal than if the transmission had been unmodulated.

Thus this unit is entirely concerned with ways of measuring, describing and finally altering the spectrum of a signal in order to achieve an acceptable signal-to-noise ratio in an instrumentation system.

The continuous spectrum

The idea of the spectrum of a periodic signal has been described in Units 8/9/10. You should recall that a periodic signal can be regarded as the sum of a set of sinusoids. The lowest-frequency sinusoid is called the fundamental and it has the same period as the periodic signal. All the other sinusoids have frequencies which are multiples of this fundamental frequency, and are called the second harmonic, the third harmonic and so on. The amplitudes and phases of the sinusoids are such that, if they were added together, they would reproduce precisely the original periodic signal.

This description of a signal in terms of its frequency components can also be used for signals which are not periodic. Such signals are termed *aperiodic*. Since most instrumentation and noise signals are aperiodic, a study of their spectra is more important to us than is the study of the spectra of periodic signals.

aperiodic

Two segments of aperiodic waveforms are shown in Figure 1. Figure 1(a) shows a waveform which is deterministic. The waveform shown in Figure 1(b) is continuous and random. Our problem here is to consider how to specify the spectrum of such waveforms.

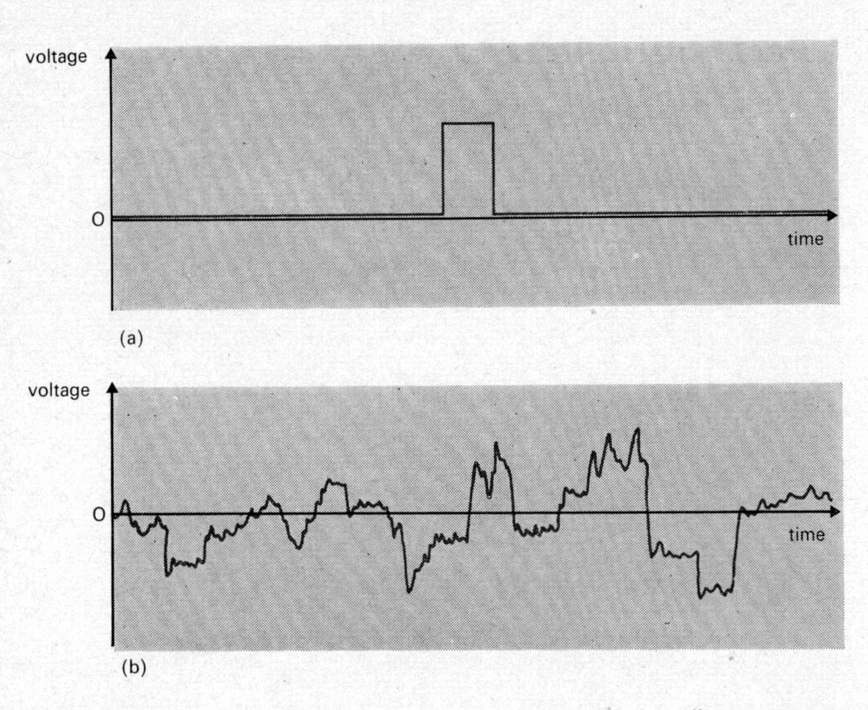

Figure 1 Two types of aperiodic waveform: (a) a single pulse type; (b) a continuous random waveform

In order to describe the spectrum of an observed waveform one must realize that one can only *know* about that finite segment of a waveform. To describe a waveform of limited duration in terms of its frequency components, Fourier's theorem, introduced in Units 8/9/10 for periodic waveforms, must be applied to aperiodic waveforms.

Fourier stated that *any* waveform which is observed for a finite length of time T_o can be fully specified by a Fourier series. That is,

$$f(t)_{\text{between 0 and } T_o} = a_0 + a_1 \sin(\omega t + \varphi_1) + a_2 \sin(\omega t + \varphi_2) + \ldots$$

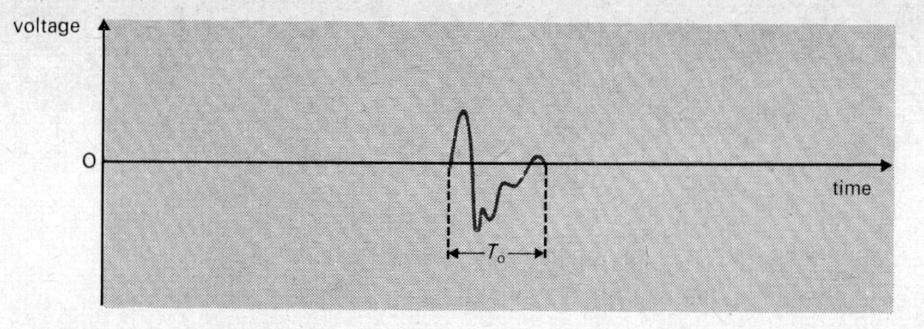

Figure 2 A continuous signal observed for time T_o

So an aperiodic waveform such as that shown in Figure 2 can be specified by its Fourier components in terms of their amplitudes and phases.

The point to recognize is that the waveform of Figure 2 is identical to one period of the periodic waveform shown in Figure 3. Thus the Fourier series which describes that periodic waveform also describes the aperiodic waveform for the observation time T_o.

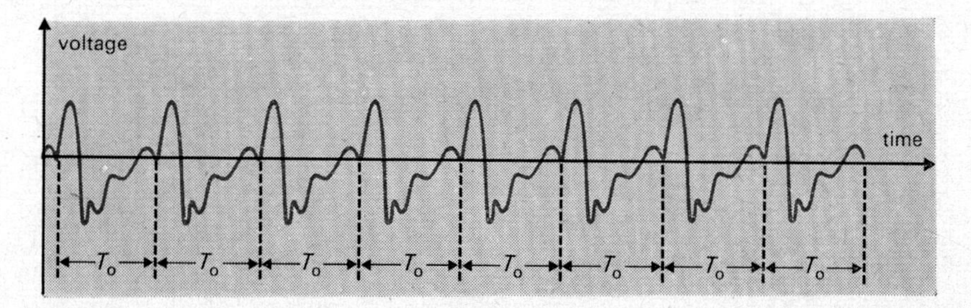

Figure 3 A periodic waveform which is identical to the waveform shown in Figure 2 during the observation interval

One way in which the Fourier components of a waveform can be represented is by plotting their amplitude against frequency on one graph and their phase against frequency on another graph.

Consider the waveform shown in Figure 4(a). Its Fourier components are described by the graphs shown in Figures 4(b) and (c). Figure 4(b) describes the amplitude of the components and Figure 4(c) their phases. If, as is true in this example, the phases of the Fourier components are 0 or π, it is possible to represent the Fourier components of a waveform on one graph, as shown in Figure 4(c). Here the amplitudes of the Fourier components are indicated by the length of the lines and the phase of the components by the sign of the lines (positive$\equiv$0, negative$\equiv\pi$).

A diagram which represents the Fourier components of a signal is known as a *spectral diagram*, or just the *spectrum*, of the waveform. If the spectrum of the waveform consists only of lines, it is termed a *line spectrum*.

spectral diagram, spectrum

line spectrum

Let us take an example of an isolated pulse which is observed for different observation intervals. Figure 5 shows the observed waveforms and their associated Fourier series.

> Why do the frequency components in the spectrum become more numerous and their amplitudes become less as T_o is increased?

As explained before, these waveforms are identical to one period of a corresponding periodic waveform of fundamental frequency $1/T_o$. All the other Fourier components of the periodic signal are harmonics of this fundamental frequency. Thus as T_o increases the frequency separation

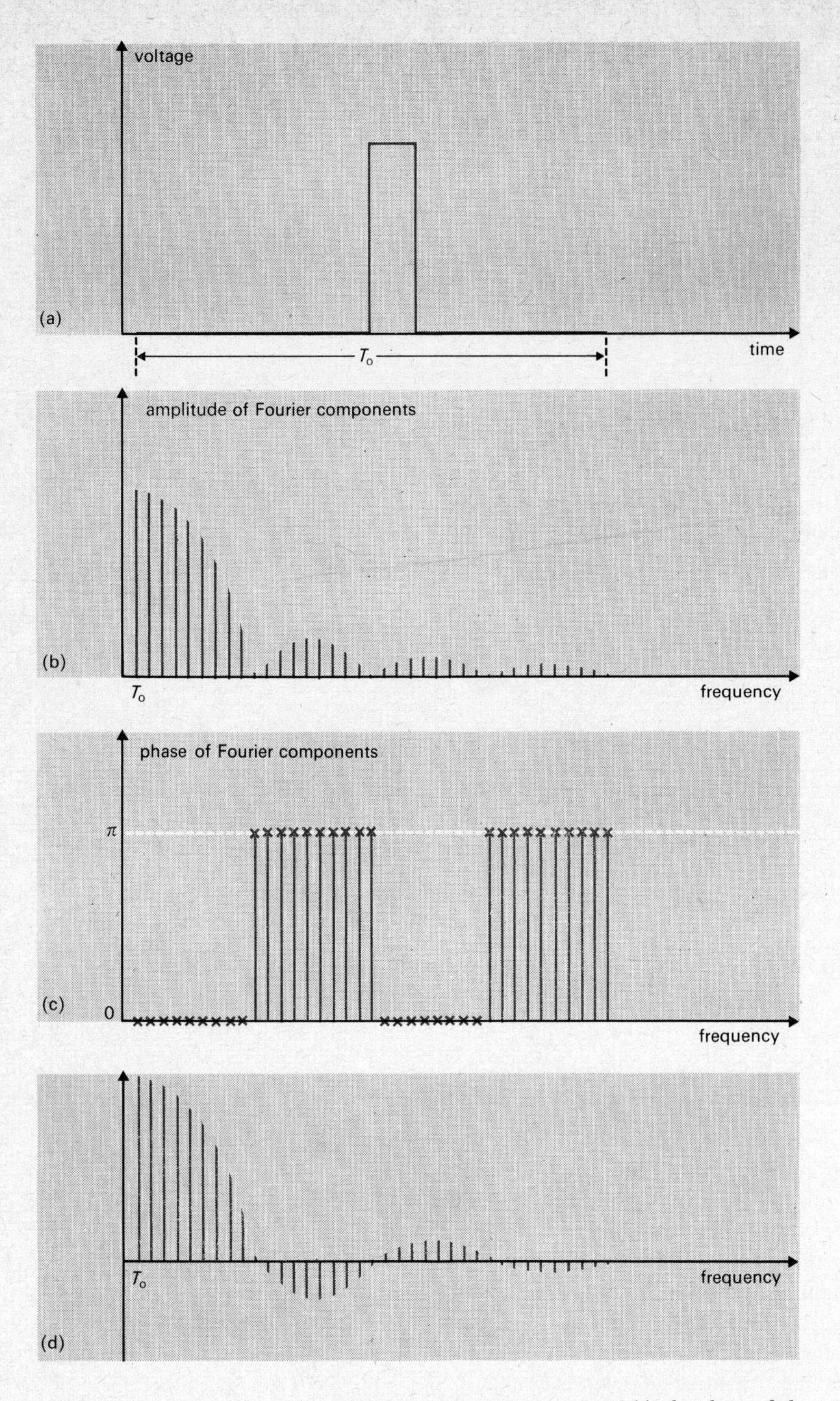

Figure 4 (b) shows the amplitude of the Fourier components and (c) the phase of the Fourier components of the waveform shown in (a). In (d) the amplitude and *phase of these Fourier components are represented on one graph*

between harmonics decreases. Now, why does the amplitude of the spectral component decrease as T_o increases?

Remember that the spectral components, when combined in their correct phases, will reproduce the original waveform. In this case the pulse shape does not alter as T_o increases. Therefore, as more and more sinusoids are required to represent the waveform, their individual amplitudes must decrease so as to keep the total voltage they produce the same as that of the pulse.

Notice that the shape of the envelope of the spectrum does not change. Evidently this is a property of the aperiodic pulse itself and not the observation time of the waveform.

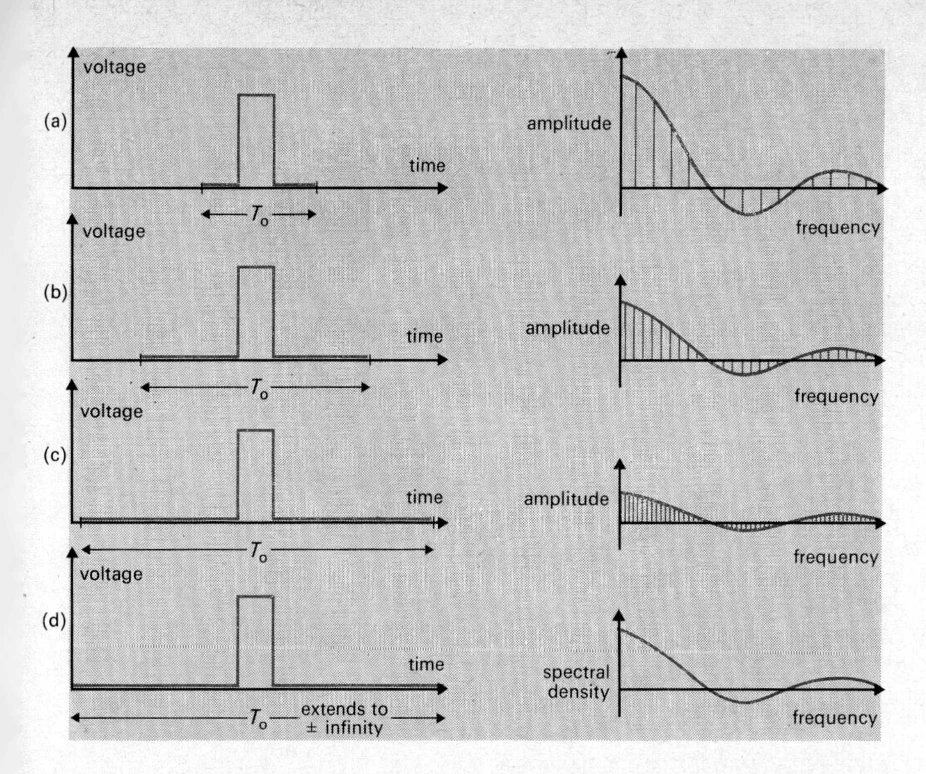

Figure 5 Spectra of a rectangular pulse for longer and longer observation time T_o

If the observation time were to extend indefinitely, the separation between the lines would become infinitesimal, the amplitudes of the sinusoids would be infinitely small and there would be an infinite number of them. The spectrum is then said to be *continuous* and, because each frequency component is infinitesimal, the spectrum is described in terms of the *spectral density* of its components. I shall be explaining spectral density in a moment, but meanwhile it is interesting to examine the spectral density of the spectrum of an isolated pulse. Figure 5(d) shows such a spectrum. The shape of the spectrum is the same as the envelope of the spectral lines, mentioned earlier.

spectral density

For a continuous random signal the spectrum does not approach a limiting case as the observation time is increased.

In addition, the envelope of the line spectrum of a finite segment of a random signal is not characteristic of the signal. Figure 6 illustrates this point. Figures 6(a), (b) and (c) show three segments of equal length of the same random signal, which have been observed at three different time intervals. Because of the random nature of the signal, the three observed segments are different and the line spectra drawn alongside the signal segments are also different. Notice that these line spectra only show the amplitudes of the Fourier components of the line spectra of the waveforms. Further graphs showing the phase of each component would be required to specify fully the signal in the observed time intervals. If we were to examine these graphs showing phase, we would find that these were different for each segment of the random signal.

Thus, although Fourier's theorem enables us to specify a segment of a random signal in terms of the amplitudes and phases of a set of sinusoids, using it on a finite segment of a random signal does not provide a spectrum which is typical of the frequency content of that random signal.

A spectrum which does characterize the frequency content of a random signal is the *power density spectrum*. The power density spectrum describes the power of the frequency components of the signal but not their phases. We can obtain an *estimate* of this spectrum, for the random signal used in

power density spectrum

Figure 6, from the line spectra describing the segments of the random signal, provided we are able to assume the signal is stationary. The first step is to find the power contained in corresponding frequency components.

For example, let us consider the corresponding components marked a_1, a_2, a_3 on the three line spectra shown in Figure 6. We can see that the amplitudes of these components are different in the three spectra and in addition their phases will be different. However, the power that one of these components can dissipate in a resistor does not depend on its phase. This power depends only on the square of the amplitude of the component and on the value of the resistor.

In particular, if the resistor has a value R, this power P is given by

$$P=\frac{V^2}{R},$$

where V is the r.m.s. value of the sinusoidal frequency component. For a sinusoid, the amplitude equals $\sqrt{2}$ times the r.m.s. value, so

$$P=\frac{A^2}{2R},$$

where A is the amplitude of the frequency component.

The next step in obtaining an estimate of the power density spectrum is to average the three component powers. This is done by adding the squares of the amplitudes and dividing by the number of spectra.

In this example we are using three spectra and the average power at the chosen component frequency is given by

$$\text{Average power}=\frac{A_1^2+A_2^2+A_3^2}{3\times 2R}.$$

This average power is then divided by the size of the frequency band between the lines to get an estimate of the power *density* at that particular component frequency.

This procedure is then performed at each of the component frequencies. The points are marked on a graph against frequency and a smooth curve is drawn through these points. The resulting curve is an estimate of the power density spectrum of the signal. Figure 6(d) shows this resulting graph. The units of power density are watts per hertz.

Strictly speaking, to interpret fully the power density spectrum of a signal, the value of the load resistor should be known. However, if the shape of the power density graph is all that interests us, we do not need to know the value of this resistor.

Another way of obtaining an estimate of the power density spectrum of a signal involves grouping together frequency components in a line spectrum and finding the average power in this group. Figure 7 illustrates this method of obtaining the power density spectrum. In this diagram a group of five frequency components is used. The average power that this group would dissipate in a load is found as before by summing the squares of the amplitudes of the components and dividing by the number of components in the group. The value obtained for power density at the centre frequency of the group is this average power divided by the bandwidth occupied by the group. If this procedure is continued throughout the spectrum, a graph of power density can be drawn as shown in Figure 7(b).

To obtain as good an estimate of the power density spectrum of the signal using this method, as compared to the previous method, the observation time must be equal to the *total* observation time of the previous method. In our particular example, the observation time must be three times as long as the previous individual observation times.

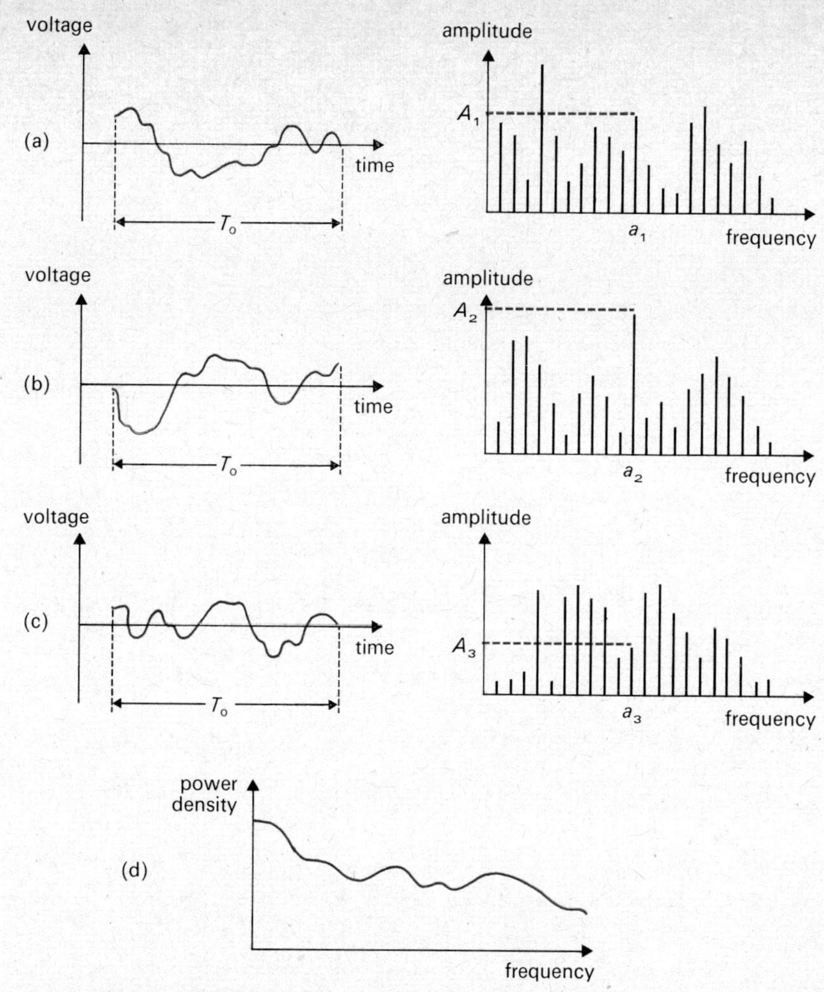

Figure 6(a), (b) and (c) show three segments of equal length of the same random signal and their corresponding frequency components; (d) shows the power density graph that can be estimated from spectra such as these

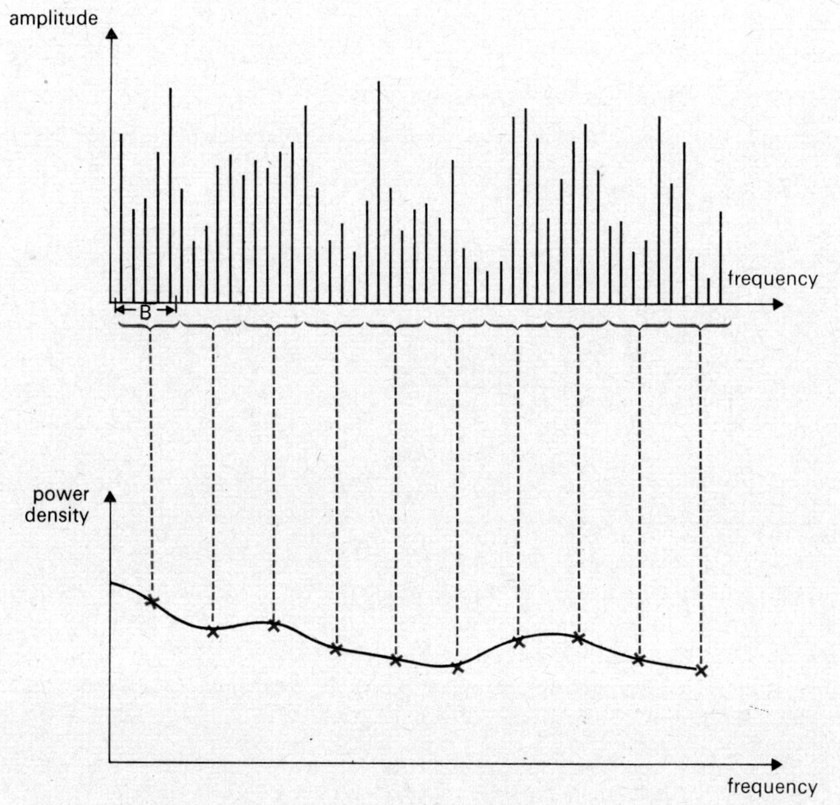

Figure 7 (a) The line spectrum of a segment of a random signal. (b) The power density spectrum of the signal, which is obtained by finding the power within the groups indicated in (a)

Electronic instruments which measure a signal's power density spectrum are called *spectrum analysers*. There are various types of these instruments, but I only need to talk in detail about one of them. This type of spectrum analyser uses several *narrow band-pass filters*, each one passing a different band of frequencies.

spectrum analysers

What is a narrow band-pass filter?

You should recall from Unit 11 that a *band-pass filter* is a circuit which only allows signal frequencies within a certain range to pass through it relatively unaffected. Signal frequencies outside this range are attenuated. A typical frequency response for a band-pass filter is shown in Figure 8. *Narrow band* refers to the range of frequencies passed by the filter (i.e. f_2-f_1) and this is said to be narrow if it is small compared with the central frequency f_0 passed by the filter.

band-pass filter

narrow band

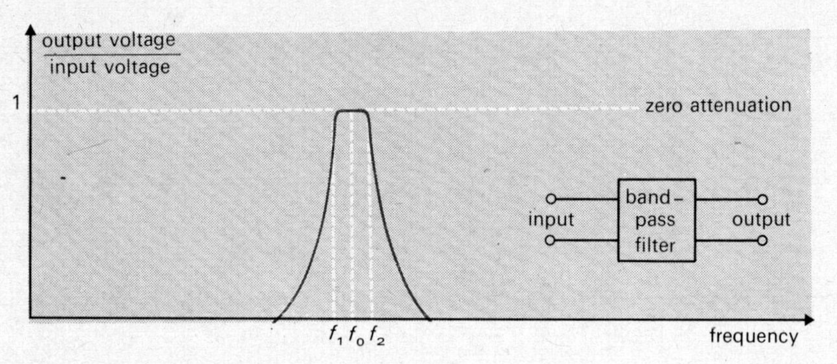

Figure 8 Frequency response of a band-pass filter

Typical frequency response characteristics of the spectrum analyser's set of filters can be seen in Figure 9. Each filter passes a different band of frequencies, but, inevitably, the filters are constructed so that there is some overlap between their pass bands. In this case the overlap is controlled so that the spectral components in the overlapping regions tend to affect two adjacent filters equally. Because of this overlap the instrument is called a *contiguous-band spectrum analyser*.

contiguous-band spectrum analyser

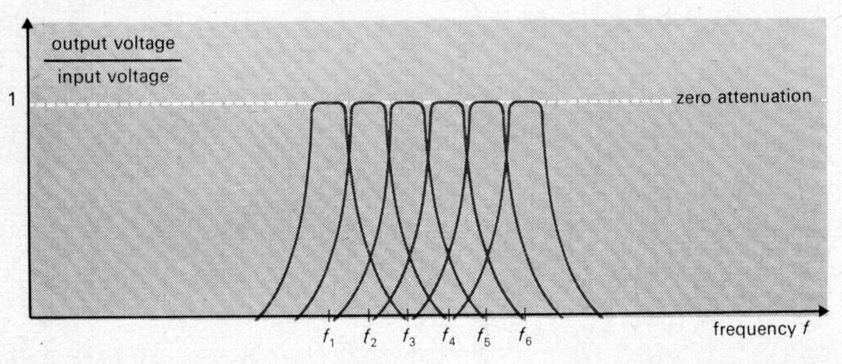

Figure 9 Frequency-response characteristics for the set of filters of the spectrum analyser

Each filter in the analyser is connected to an electronic voltmeter which indicates the r.m.s. value of a narrow-band part of the signal fed to it via the filter. Thus the signal is divided into a number of parallel, narrow-bandwidth continuous signals. The length of the observation time is the period over which the r.m.s. values are averaged in the spectrum analyser. Figure 10 shows a schematic diagram of the analyser.

Each indicated r.m.s. value is, of course, the r.m.s. value of the frequency components in a particular part of the spectrum of the waveform, so the r.m.s. values can be plotted against their corresponding filter's centre frequency.

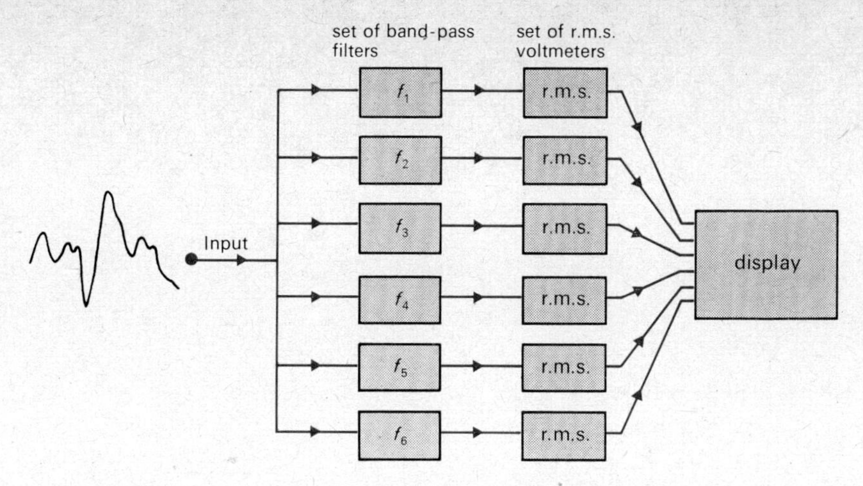

Figure 10 A schematic diagram of a contiguous-band spectrum analyser

A possible plot is shown in black in Figure 11. The frequency axis is marked out in the cut-off frequency of the filters. The vertical axis shows the r.m.s. voltages indicated by the voltmeters. Thus each point indicates the voltage within a band. The band for the first point is f_1–f_2, for the second point it is f_2–f_3.

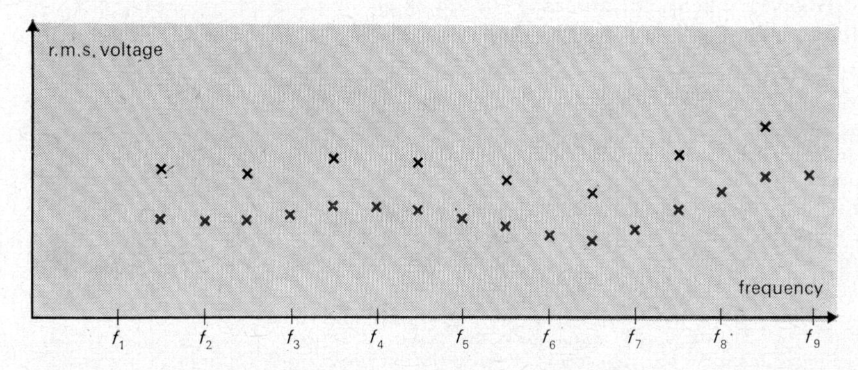

Figure 11 Two possible plots of the spectrum of a signal as measured by two different contiguous-band spectrum analysers. The red crosses are the results from the analyser with the smaller bandwidth

> If we had taken a different contiguous-band wave analyser with filters of half the bandwidth, so that there were twice as many of them, what would the plot have been like?

Each filter would pass half the proportion of the signal's spectrum passed by the previous spectrum analyser's filters. So the power associated with a filter's output would drop by a half, and since this power is proportional to (r.m.s. voltage)^2, the r.m.s. voltages would drop by $1/\sqrt{2}$ on average.

The plot for this spectrum analyser is shown in red in Figure 11. The points are $\sqrt{2}$ less than for the first spectrum analyser and there are twice as many of them.

Now I hope you can see that dividing the measured voltage by the square root of the bandwidth of the filter will result in the same graph from each spectrum analyser, as shown in Figure 12.

The units on the vertical axis of this graph will be r.m.s. volts per hertz$^{\frac{1}{2}}$. This is, of course, equal to the square root of the units for power density (assuming a 1 Ω load). So the graph is a plot of the square root of power density against frequency. In fact, this is a common way of plotting the frequency characteristics of a signal, although the normal method of

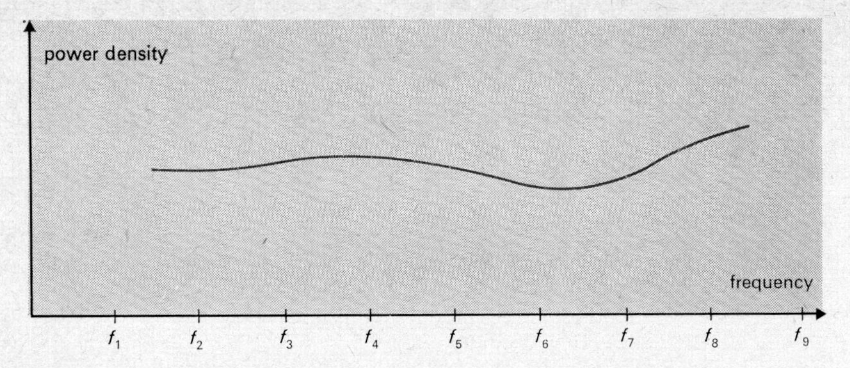

Figure 12 A plot of the power density spectrum of the signal in units of volts per root hertz

plotting power density against frequency is also often used. For the remainder of these units I shall plot the power density spectrum as power density against frequency.

SAQ 4

SAQ 4

If the r.m.s. value of the output of one of the filters is 8 V when applied to a 50 Ω load and the bandwidth of the filter is 40 Hz, what is the measured power density?

normalization

Figure 13 compares the power density spectra of a number of pulses of different duration. The graphs have been multiplied by a scaling factor so that the maximum value always equals one. This is called *normalization*. As the pulse length is shortened its power density spectrum becomes more uniform over a wider range. I shall be referring to this again later in this unit.

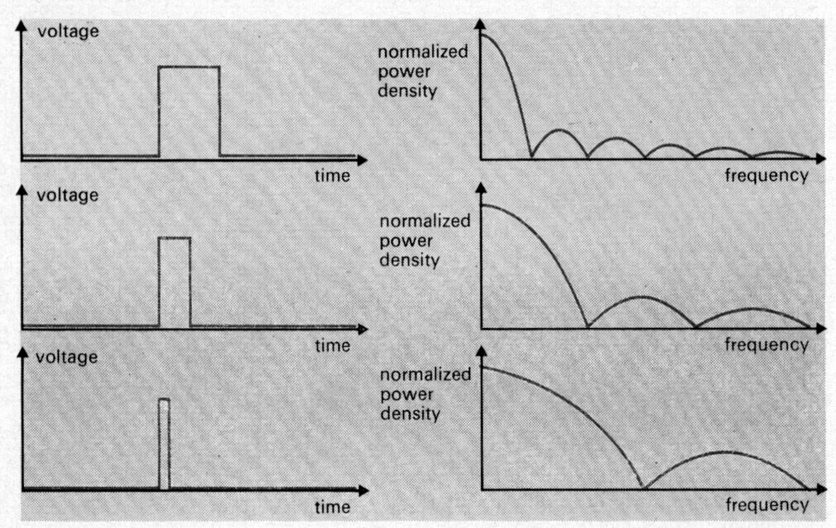

Figure 13 The effect that the duration of a pulse has upon its power density spectrum

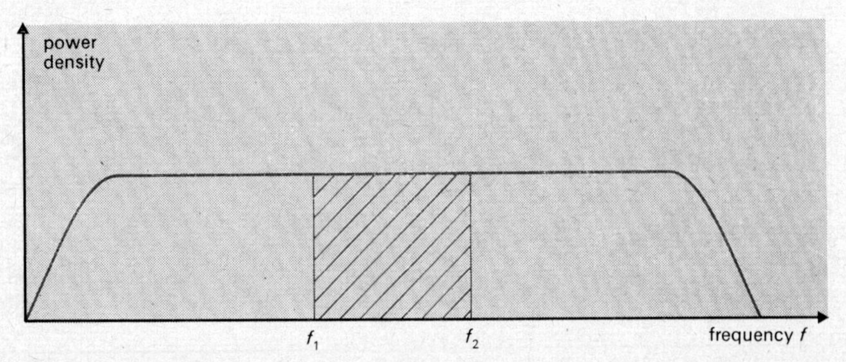

Figure 14 A uniform power density spectrum; the power in the restricted bandwidth is given by the shaded area

It is important to understand the use of the term 'density' in the power density spectrum, so let us go over it again. Suppose we are studying a signal with a power density which is uniform over most of its spectrum, as shown in Figure 14. We apply the signal to the input of a spectrum analyser of the type described above, and obtain a graph of voltage against frequency. We find that the r.m.s. value of the output of each filter is the same because the filters have equal bandwidths.

But what voltage will this be?

As explained earlier, this voltage will depend upon the bandwidths of the filters. If they have a 50 Hz bandwidth, each one will pass spectral components with a total of ten times more power than if they had a 5 Hz bandwidth. Then, since power is proportional to (r.m.s. voltage)^2, the voltmeter will read $\sqrt{10}$ moıe.

Look again at Figure 14. You will see that, although the power density spectrum is uniform over a wide range of frequencies, at low and high frequencies the power density falls to zero.

What do you suppose the area under the graph of the power density spectrum represents?

The area has the dimensions of power – that is, power density (power/frequency) multiplied by frequency – and the area under the graph represents the total power in the signal. If the signal were restricted so that it only contained frequencies in the range f_2–f_1, as shown in Figure 14, the power in this restricted bandwidth signal would be given by the shaded area marked on the graph.

SAQ 5

SAQ 5

Explain how you would characterize the probability density function of a Gaussian noise waveform which has zero mean and a power density spectrum such as that shown in Figure 14.

Noise signals with commonly occurring power density spectral shapes are given special names. For instance, noise which has a uniform power density spectrum over a range of frequencies greater than the range which is of immediate interest is called *white noise*, by analogy with white light. White light can be thought of as having a uniform power density spectrum over the visible range of frequencies.

white noise

Figure 15 shows some power density spectra of noise.

2.1 Some other methods of measuring the spectrum of a signal

Finally we should note that a common form of spectrum analyser has only one filter and one voltmeter. Instead of having a set of filters,

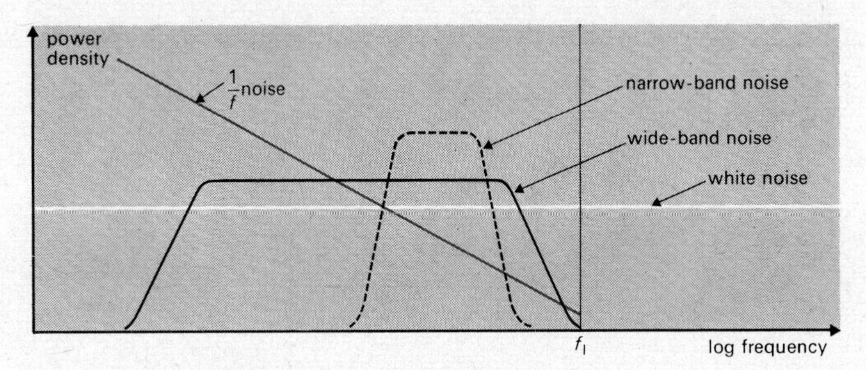

Figure 15 Power density spectra of different kinds of noise

with each one 'looking' at a different frequency band, this type uses just one narrow band-pass filter. The instrument is used to analyse a spectrum by varying the centre frequency of the filter so that it passes different parts of the spectrum in turn. If the centre frequency is varied manually and the output reading is displayed on a meter, the instrument is called a *wave analyser*. If the filter centre frequency is varied electronically and the output is displayed on a cathode ray tube, the instrument is called a *spectrum analyser*. A spectrum analyser of this type is used in the television programme 'Modulation and noise'. When using this type of spectrum analyser you have to be able to assume that the spectrum of the signal does not change faster than the time taken for the analyser to examine the entire spectrum of the signal. The r.m.s. output of the filter at different frequencies is plotted against frequency to give the power density spectrum.

wave analyser

spectrum analyser

Neither type of spectrum analyser so far described is capable of obtaining phase information relating to a signal. This can be obtained for a segment of a signal of duration T_0 by calculation on the quantified segment.

First the waveform is sampled at greater than twice the maximum frequency present in the signal. Now, as we know, these samples fully describe the signal, so it is possible to use these samples to calculate a Fourier-component representation of the segment. Standard computer programs for doing this exist. An ingenious algorithm for doing this calculation quickly on a computer is known as the fast Fourier-transform algorithm.

2.2 Summary of Section 2

In this section I have explained that a segment of an aperiodic signal can be fully specified by its Fourier components in a line spectrum. This is just one graph if the phases of Fourier components can only be 0 or π. If, however, the Fourier components have other values of phase, separate graphs are required to specify the amplitudes and phases of the components.

The line spectrum of a continuous random signal does not approach a limit as the observation time interval is increased. So, for this type of signal, the frequency content of the signal is described in terms of its power density spectrum. This power density spectrum does approach a limiting shape as the observation time is increased.

The area under a power density graph is equal to the total power of the signal.

Instruments which measure power density spectra are called spectrum analysers. There are many types of these instruments. One major type consists of a tunable filter and a voltmeter, while another type consists of a contiguous band of non-tunable filters, each of which is connected to a voltmeter.

Filter characteristics

In Unit 11 I drew the frequency response of a low-pass filter without marking the axes of the graph. I did this because, at that stage, it was only important for you to appreciate the idea of the low-pass filter circuit. But it is often necessary to obtain, from the graph of the frequency response of a filter, values for the attenuation that the filter will produce at a certain frequency. Therefore I want to talk briefly about the way in which the response of a filter is usually plotted.

The frequency response of a filter is sometimes called the characteristic of the filter. The characteristic of a filter shows how the filter affects the amplitude *and phase* of sinusoids of different frequencies.

Now, because the characteristic of the filter may have to be plotted over a wide range of frequencies, it is convenient to use a *logarithmic frequency scale.* This is most easily done by using graph paper which has a *logarithmic scale.*

logarithmic frequency scale

On a logarithmic scale the space allotted to each *decade* of frequency is the same. A decade is a ten-to-one frequency range. Thus the decade 10–100 Hz has the same space as the 100–1000 Hz and 1–10 kHz ranges.

decade

In addition it is conventional to plot the other variable in the amplitude response, the output voltage, in terms of a logarithmic unit called the *decibel* (dB).

decibel (dB)

If $\hat{V}_{in}$ is the peak value of a sinusoid at the input to a filter and $\hat{V}_{out}$ is the peak value of the sinusoid at the output of the filter, then the *ratio* of these two voltages may be expressed in decibels as

$$20 \log_{10} \frac{\hat{V}_{out}}{\hat{V}_{in}}.$$

Alternatively, r.m.s. values may be used, because

$$\text{Peak value} = \text{r.m.s. value} \times \sqrt{2}$$

and the $\sqrt{2}$ factor cancels out in the ratio of the output and input voltages.

What is the ratio of output to input voltage expressed in decibels if:

(a) $V_{in} = 10$ V and $V_{out} = 1$ V;

(b) $V_{in} = 4$ V and $V_{out} = 2$ V?

(a) $20 \log_{10} \frac{1}{10}$ dB $= -20 \log_{10} 10$ dB $= -20$ dB.

(b) $20 \log_{10} \frac{2}{4}$ dB $= -20 \log_{10} 2$ dB $= -6$ dB.

SAQ 6

SAQ 6

What is the value of V_{out}/V_{in} in decibels, when the power that a sinusoid can produce in a resistor of resistance R at the output of a filter is one-half the power the sinusoid could have produced in the same resistor before it passed through the filter?

The frequency response of one type of low-pass filter is shown in Figure 16.

The *cut-off frequency* of the filter is the frequency at which

$$\frac{V_{out}}{V_{in}} = -3 \text{ dB}.$$

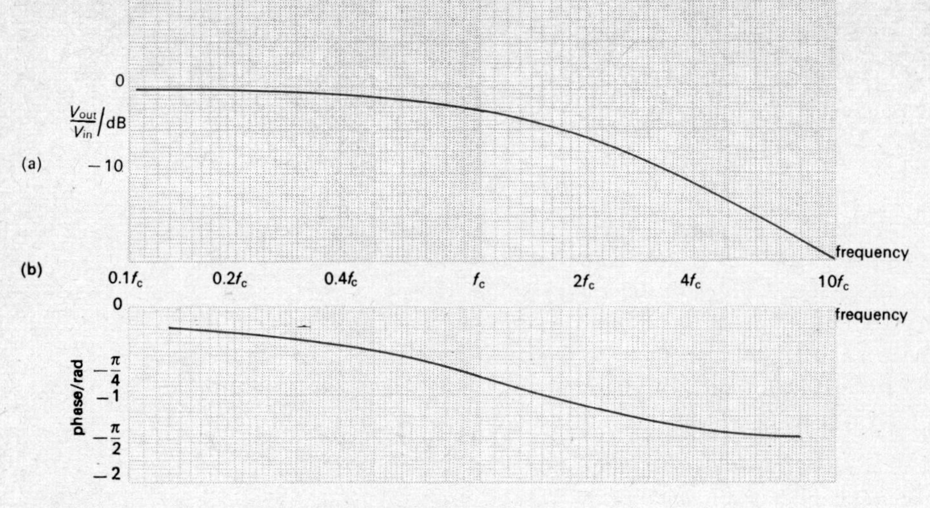

Figure 16   Frequency response of a low-pass filter

It is possible to construct many other types of filter. The characteristics of three other types are shown in Figures 17, 18 and 19.

From the characteristics of a *high-pass filter* we can see that it attenuates frequencies below its cut-off frequency and has relatively little effect upon those frequencies above its cut-off frequency.

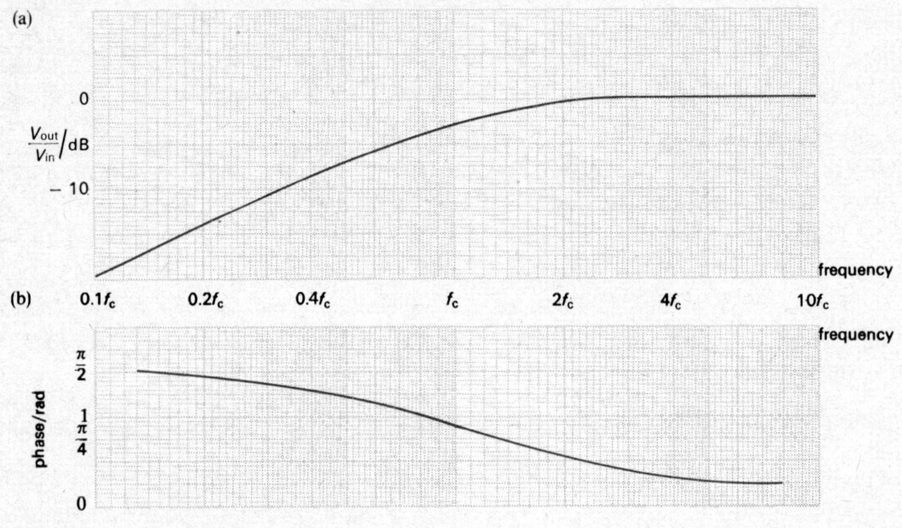

Figure 17   Frequency response of a high-pass filter

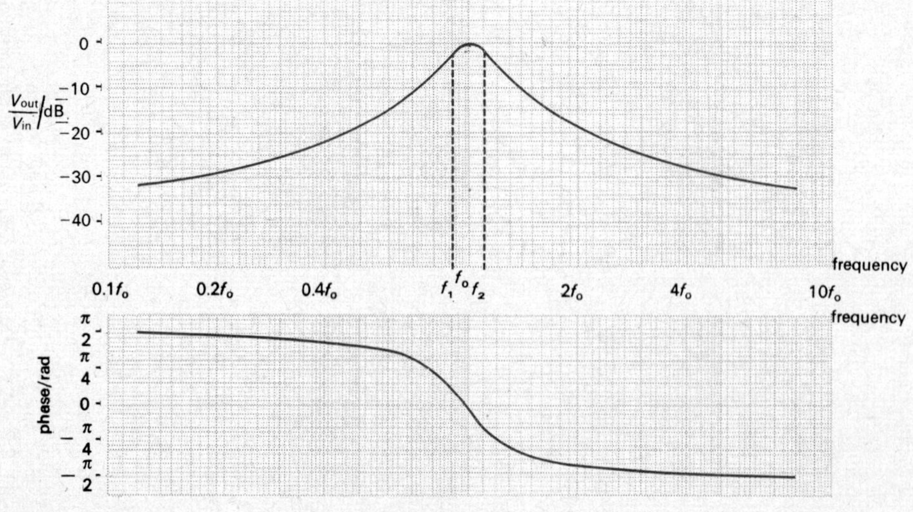

Figure 18   Frequency response of a band-pass filter

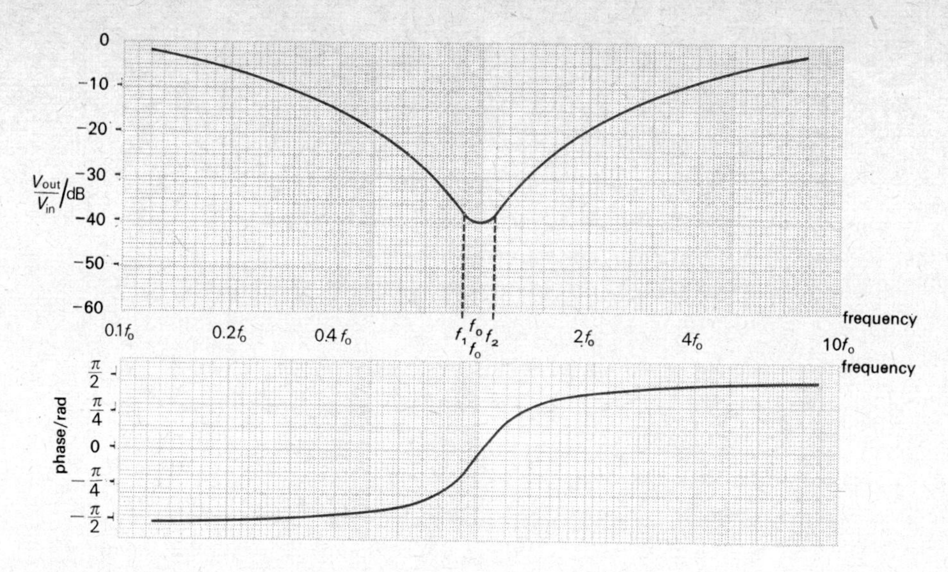

Figure 19 Frequency response of a band-stop filter

SAQ 7

SAQ 7

Describe the characteristics of the band-pass and band-stop types of filter shown in Figures 18 and 19.

3.1 Signal filtering

Of all the methods available for separating signals from noise the most widely used is that involving filtering. If the message signal of interest has a different spectrum from the noise, it is always possible to design a filter which will attenuate the noise more than it will attenuate the signal. In other words, filtering can be made to improve the *signal-to-noise ratio.*

In the television and radio programmes associated with this unit you will see and hear examples of the use of filtering to improve signal-to-noise ratios.

I mentioned in Unit 11 that to minimize the effect of noise interfering with a signal you must arrange the bandwidth of the instrumentation system to be exactly the same as the bandwidth of the signal.

Let us now consider some examples of this.

1 I have explained one example in Unit 11. It was the example describing the tuning of a radio (or television) receiver.

In the receiver a band-pass filter is used to equate the bandwidth of the receiver to that of the signal. If the bandwidth of the receiver is too large, extra noise or interference from other stations will be heard. If the bandwidth of the receiver is too small, some of the signal will be lost.

2 The demonstration of a variable-reluctance pressure transducer shown in the television programme 'Pressure transducers' provides another example of the use of a filter to match the bandwidth of the system to that of the signal. There the output of an a.c. bridge had to be amplified to make it suitable to drive a phase-sensitive circuit and to be displayed on an oscilloscope.

The frequency of the bridge energizing signal was 1 kHz. The only changes of pressure to be shown were ones taking place over a few seconds. This type of variation meant that the output signal from the bridge had a spectrum which consisted of a narrow band of frequencies around 1 kHz.

(The reason why is discussed in section 5.1.) The amplifier was matched to this spectrum by designing it to act like a band-pass filter. It amplified frequencies around 1 kHz and did not amplify frequencies outside this region. In this way it was possible to prevent significant interference from the cameras and other equipment in the studio.

3 The fuel gauge in a car is also matched to the signal it has to display. It displays the average level of petrol in the tank while ignoring the rapid fluctuations in level caused by the motion of the car.

The fuel instrumentation system is designed to act like a low-pass filter, allowing slow variations in the level of the fuel to be displayed while ignoring the rapid fluctuations.

I am now going to describe two other examples of filtering, in which, unlike the previous examples, the interference occupies some or all of the frequency spectrum of the signal.

One example of this use of filtering can be found in some hi-fi record players. The circuit doing the filtering is termed a *scratch filter*. Its name is derived from the fact that it is designed to reduce the effect of scratches on a *badly worn* record. The filter is actually a low-pass filter with a cut-off frequency at about 10 kHz. The reason why it is effective in reducing the noise caused by the scratches is that the frequency spectrum of the noise due to the scratches and the spectrum of the signal are different.

scratch filter

Many recorded signals of speech and music will have a power density spectrum which decreases towards the upper limit of the audio-frequency range. The shape of the noise power density depends upon the characteristics of the pick-up. Figure 20(a) shows fairly typical signal and noise spectra. The effect that the scratch filter has on these two spectra is shown in Figure 20(b).

The area under the power density spectrum of the noise is reduced by about one-third. This corresponds to a reduction in noise power of about one-third. The area under the spectrum of the signal is not

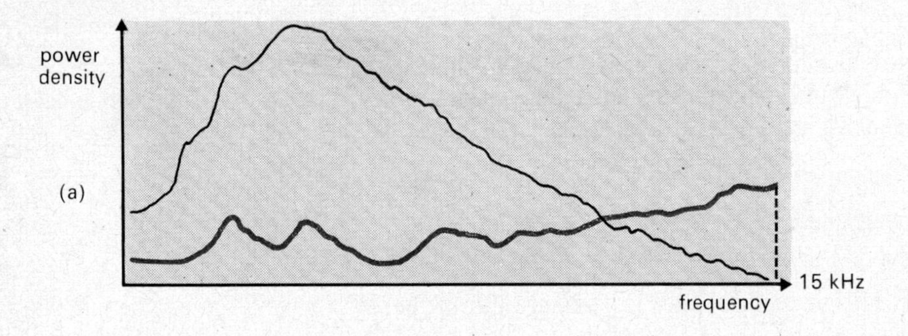

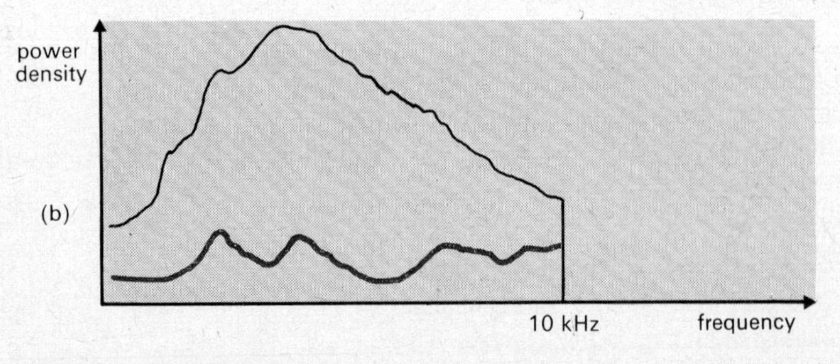

Figure 20 The power density spectra of the signal and the noise due to scratches (a) before filtering and (b) after filtering

affected nearly so much. Therefore, the signal-to-noise power ratio increases when the scratch filter is used. Most listeners consider this to be accompanied by an improvement in the quality of the sound being reproduced, even though some of the higher-frequency notes are missing.

If no scratches are present on the record, the scratch filter should not be used. Can you see why?

In this case the only effect the filter would have is to remove components in the wanted signal which are above 10 kHz. The sound quality is impaired somewhat and there is no corresponding reduction in noise.

The second case of in-band interference I want to describe occurs in many instrumentation systems. It is interference from the a.c. mains supply. The mains voltage can cause 50 Hz interference to be added to instrumentation signals. This is especially likely if a mains cable and an instrumentation cable are in close proximity.

Assume this 50 Hz interference is inside the frequency spectrum of the signal, as shown in Figure 21. What sort of filter should be used to remove the interference?

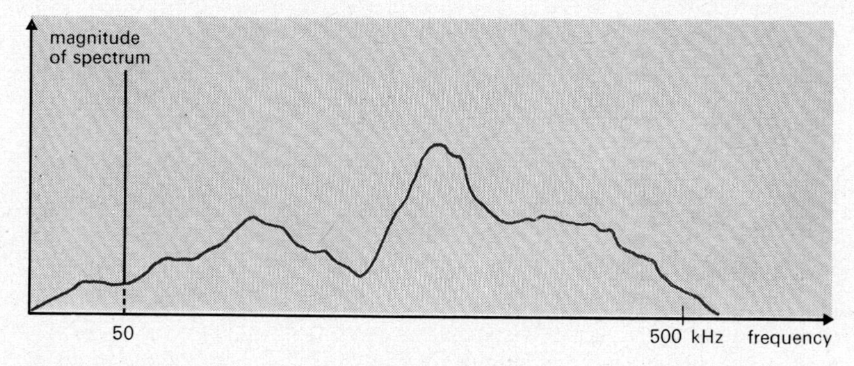

Figure 21 The spectrum of a signal with 50 Hz *interference*

A narrow band-stop filter centred on 50 Hz will remove most of the 50 Hz interference and the frequency components in the signal centred around 50 Hz, while not affecting the majority of the frequency components of the signal. But what will the filter do to the phases of the Fourier components of the signal?

To answer this question it is necessary to consider the phase response of a typical narrow band-stop filter. Figure 22 shows just such a response. The response is non-linear over the band of frequencies contained in the signal. This may affect the signal so much that it prevents the use of a band-stop filter to remove 50 Hz interference.

A very effective way of removing 50 Hz interference is to place an earthed conductor between the source of the interference and the part

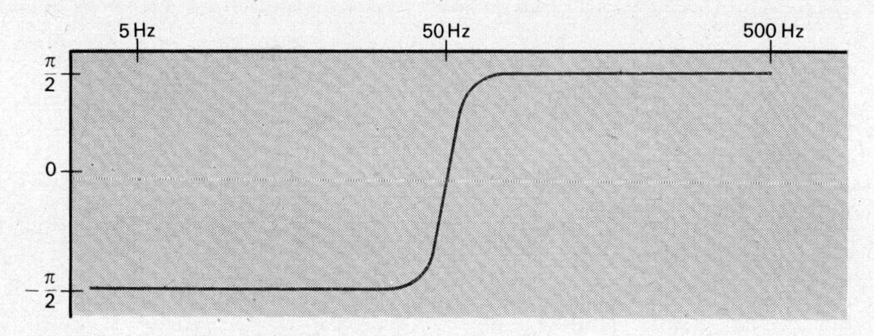

Figure 22 Phase response of a narrow band-stop filter

of the system which is picking up the interference. This prevents the 50 Hz interference being superimposed upon the signal.

This technique is referred to as *shielding*, or *screening*. It is done by placing sensitive parts of the system in earthed metal boxes and by using electrical cables surrounded by an earthed conductor, which are called *screened cables*.

shielding, screening

screened cables

Another way of decreasing the 50 Hz interference once it has been picked up involves the use of a differential amplifier. This was briefly described in Unit 11 and is considered again in Unit 16.

I have now mentioned some examples of filtering. Other examples have been mentioned or implied in earlier units and others will occur in future units.

The autocorrelation function of a signal

The *autocorrelation function* is the last of the signal parameters listed in Unit 11. It is closely related to the power density spectrum of the signal, mentioned earlier. Indeed, it is an alternative and often more convenient way of specifying the power density spectrum of a signal.

autocorrelation function

First I shall explain the function in general terms and then give some specific examples.

The process of autocorrelation establishes the relationship between the signal and a time-shifted version of itself. This relationship for all possible time shifts is described by the autocorrelation function.

This function can be obtained from samples of a signal. The value of the autocorrelation function for a delay of time t is called the *autocorrelation coefficient* of the waveform for delay t.

The instrument which measures the autocorrelation function of a signal is called an *autocorrelator*, or sometimes just a *correlator*. A schematic diagram of one type of autocorrelator is shown in Figure 23. Now, how does it work?

autocorrelator, correlator

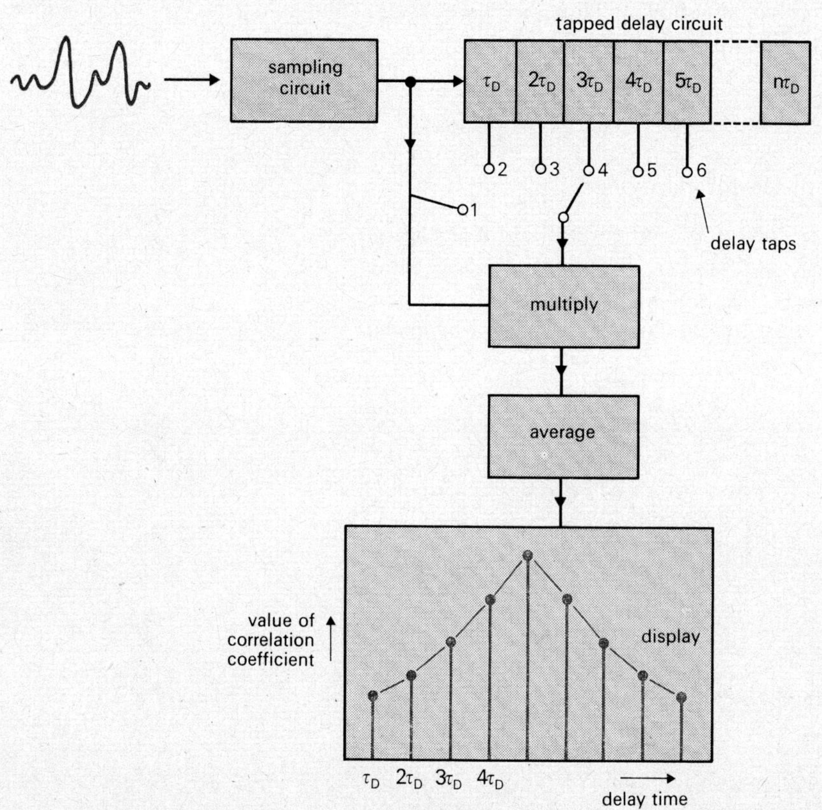

Figure 23 Schematic diagram of an autocorrelator

First the signal is sampled as was described in Unit 11. Then the samples are fed one by one into the multiplication circuit and into the tapped delay circuit. The tapped delay circuit is an electronic circuit which delays the samples by some multiple of τ_D, the time between samples, depending on which of its output taps is used. For example, if tap no. 4 is used, the delay amounts to $3\tau_D$. You will meet a similar circuit again in Unit 13, where it will be treated in more detail, but for the moment just accept this circuit as a means of delaying a series of samples before the multiplication process.

As drawn in the diagram the correlator is adjusted so that it is calculating the autocorrelation coefficient of the signal corresponding to a delay of $3\tau_D$. Sample values of the signal are being multiplied one by one with sample values that occurred $3\tau_D$ previously. The resulting products are then averaged to get a value for the autocorrelation coefficient for a delay of $3\tau_D$. In practice this process is only performed on a finite number of samples.

To obtain different autocorrelation coefficients of the signal, different delays must be used and then a graph of the autocorrelation function can be produced on the display device, as shown in Figure 23. The display device is usually an oscilloscope screen.

I shall show you later that an autocorrelator is in some instances capable of detecting the presence of a signal which is affected by noise.

Let us consider some examples of how to calculate the autocorrelation coefficients of signals.

The first example will involve only binary signals (i.e. signals with only two possible values), whose values I have chosen to be ± 1 V, since multiplication of ± 1 by ± 1 is so simple.

Example 1

Consider the square wave shown in Figure 24(a). This has only two values, either $+1$ or -1. We shall take samples of this signal ten times per cycle of the square wave, as shown. Thus, for two cycles of the square wave the sample values we obtain will be

$$+1\ +1\ +1\ +1\ +1\ -1\ -1\ -1\ -1\ -1\ +1\ +1\ +1\ +1\ +1\ -1\ -1\ -1\ -1\ -1$$

The negative values are shown in red for clarity.

The autocorrelation function is the graph of several autocorrelation coefficients, and each coefficient is obtained in essentially the same manner. To distinguish between the coefficients, I shall call the coefficient produced for zero delay the *first autocorrelation coefficient*

first autocorrelation coefficient

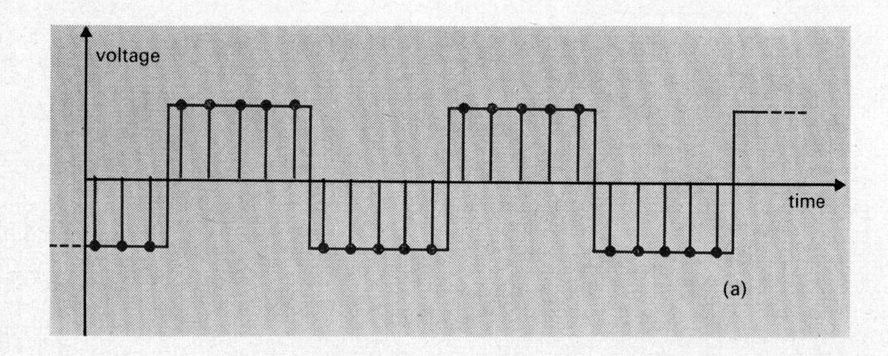

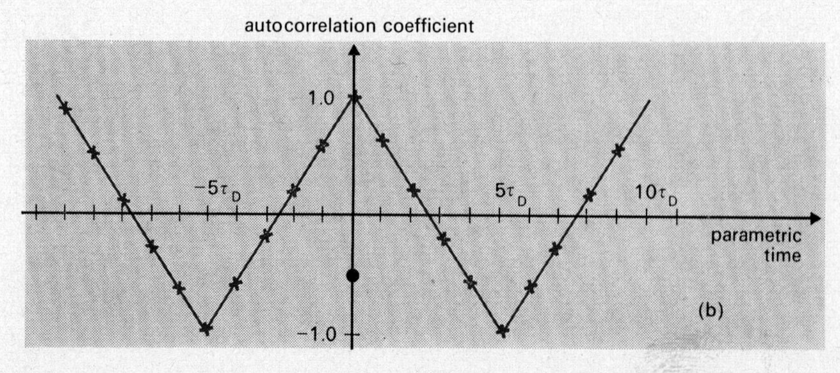

Figure 24 (a) A square wave and (b) its autocorrelation function

and the coefficients produced for delays of 1, 2, 3, 4, . . . times the time between samples the second, third, fourth, fifth, . . . autocorrelation coefficients respectively.

To obtain the first autocorrelation coefficient, you lay a duplicate version of the waveform alongside the original waveform as shown in Table 1.

Table 1 Calculation of the first autocorrelation coefficient of a waveform

waveform	+1	+1	+1	+1	+1	−1	−1	−1	−1	−1	+1	+1	+1	+1	+1	−1	−1	−1	−1	−1
duplicate	+1	+1	+1	+1	+1	−1	−1	−1	−1	−1	+1	+1	+1	+1	+1	−1	−1	−1	−1	−1
products	1	1	1	1	1	1	1	1	1	1	1	1	1	1	1	1	1	1	1	1

To obtain the coefficient you multiply each sample value by the corresponding sample value of the duplicate waveform. The result is the row labelled *products* in Table 1. In each case the value of the product is +1 in this example. You then take the average of all these products, which is evidently +1 in this case. *This is the first autocorrelation coefficient.*

To obtain the second autocorrelation coefficient, you move the duplicate version of the waveform one sample interval to the right, corresponding to a delay of one sample interval. You thus again obtain two waveforms laid alongside each other, as shown in Table 2.

Table 2 Calculation of the second autocorrelation coefficient of a waveform

waveform	+1	+1	+1	+1	+1	−1	−1	−1	−1	−1	+1	+1	+1	+1	+1	−1	−1	−1	−1	−1
one shift	−1	+1	+1	+1	+1	+1	−1	−1	−1	−1	−1	+1	+1	+1	+1	+1	−1	−1	−1	−1
products	−1	1	1	1	1	−1	1	1	1	1	−1	1	1	1	1	−1	1	1	1	1

The products obtained after shifting the waveform in this way are again shown in the third row. Now every fifth value is −1, since $+1 \times -1 = -1$. The average of all these products is now 0.6. (Check, if you like, to be quite sure.)

To obtain the third autocorrelation coefficient, the duplicate waveform is shifted a further sample interval to the right.

Write down the two sets of waveform samples with this further shift included, work out the products and calculate the new autocorrelation coefficient.

The products are this time +1, +1, +1, −1, −1, etc., whose average is 0.2.

After five steps in this process, I hope it is clear that the autocorrelation coefficient will be −1, since after five shifts to the right all the positive values of the waveform will be next to all the negative values of the duplicate waveform, and vice versa, thus each product will now be −1, and so the average will be −1.

Further shifts to the right will cause the autocorrelation coefficient to move in equal steps back towards its original value of +1.

The autocorrelation function is the graph of autocorrelation coefficients plotted against the number of shifts in time of the duplicate waveform.

For this waveform the autocorrelation function is evidently that shown in Figure 24(b), that is, a simple zig-zag between -1 and $+1$.

The time delay through which the duplicate waveform is shifted is called *parametric time*. Notice that in this example the parametric time *intervals* τ_D are the same as the time intervals between the actual waveform samples* and that the period of the autocorrelation function in parametric time is the same as the period of the original waveform.

parametric time

So, summarizing, to carry out the operation known as determining the autocorrelation function:

1 Lay the waveform and its duplicate side by side but shifted in time relative to each other by a time delay known as the *parametric time*.

2 For each chosen value of parametric time, obtain the autocorrelation coefficient, that is,

(a) obtain the product of each sampled value of the waveform and the sampled value of its delayed duplicate, which are presented to the multiplier at the same instant;

(b) take the average of all these products.

3 Plot the values of these autocorrelation coefficients against the corresponding values of parametric time.

SAQ 8

SAQ 8

Find the autocorrelation function of one period of a sine wave using the data given in Table 1 of Unit 11. (*Note*. To reduce the number of values which you need to calculate, the autocorrelation coefficients nos. 9, 10, 11 and 12 are -0.25, 0, 0.25 and 0.44. Therefore you only have to calculate the first eight coefficients.)

Example 2 The autocorrelation function of random noise

Figure 25 shows a segment of Gaussian noise that was first shown as Figure 21(a) in Unit 11. The autocorrelation function of this segment of noise can be obtained from the fifty sample values of this noise which were tabulated in Table 2 of Unit 11. This would be a tedious operation to do manually, as in Example 1, but it takes a computer very little time to do it for us. The resulting autocorrelation function is shown in Figure 26.

Notice that the autocorrelation function has a maximum value for zero delay and that for shifts of greater than one sample interval its value is very nearly zero.

These two examples illustrate some of the differences between the autocorrelation functions of periodic and random waveforms.

The autocorrelation function of a periodic waveform is itself periodic and the maximum value of correlation occurs at regular time shifts. In fact the period of the waveform is the same as the period of the autocorrelation function, provided τ_D is chosen to be equal to or less than the minimum sampling time interval allowed by the sampling theorem. Note, however, that the waveform is periodic in real time, the function is periodic in parametric time.

**It is not necessary, in determining an autocorrelation function, to make the steps in parametric time identical to the intervals between samples. However, this need not concern us here.*

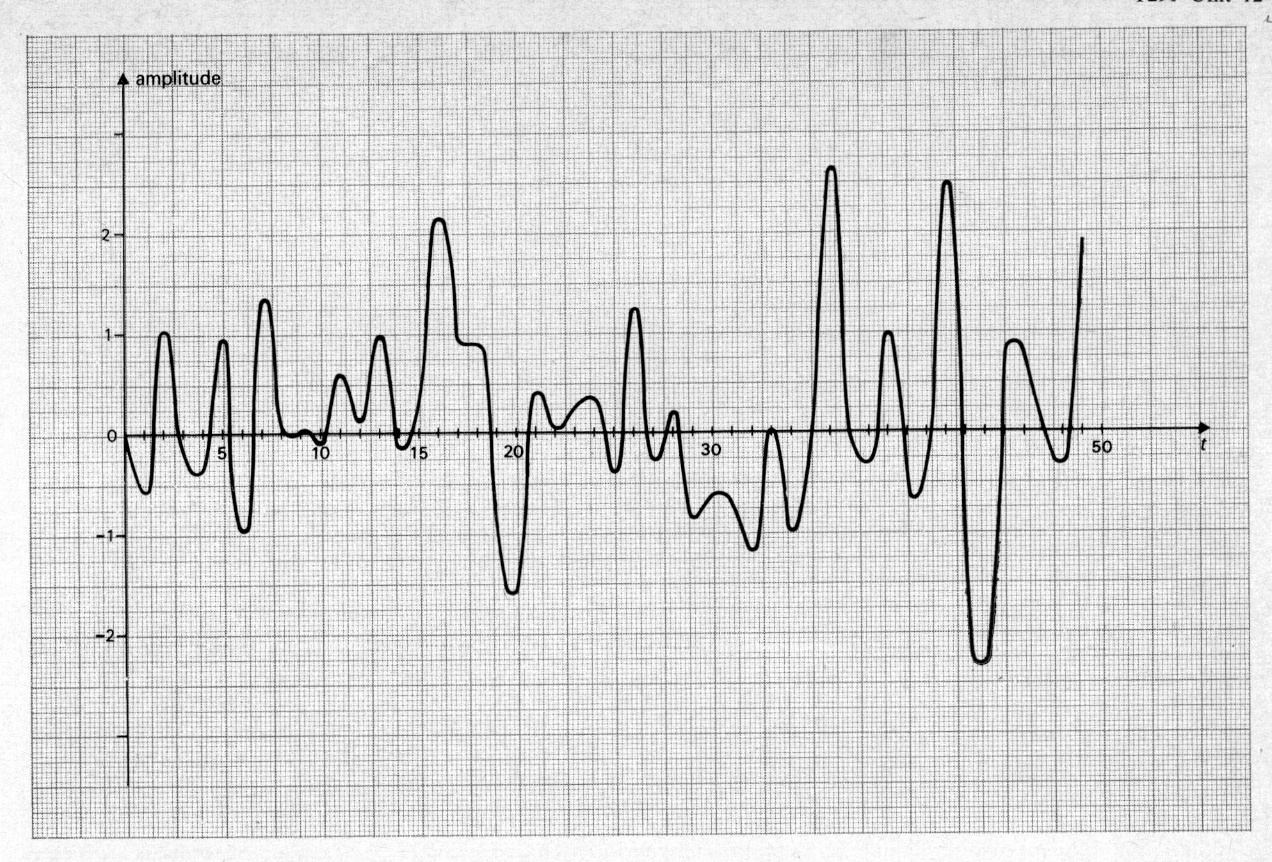

Figure 25 A segment of Gaussian noise

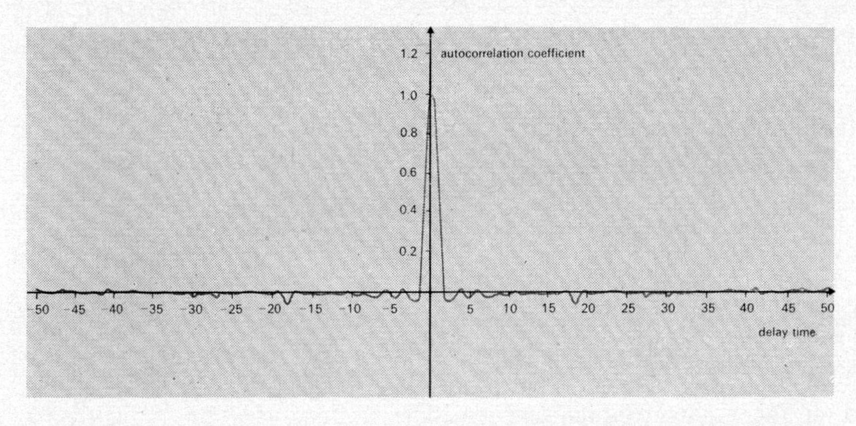

Figure 26 Autocorrelation function of the noise sample values given in Table 2 of Unit 11

On the other hand, the autocorrelation function of random noise is not periodic, it consists of a single spike centred around zero time delay.

But random noise is not the only signal which has this type of autocorrelation function. A signal which consists of a short pulse has a similar autocorrelation function. Figure 27 shows such a pulse and its autocorrelation function. Remember from Section 2 that this type of pulse has a uniform power density spectrum over a range of frequencies. The difference between the pulse and random noise with a uniform power density spectrum over the same range is in the phase relationship between their spectral components. This phase information about the spectrum is lost in finding both the power density spectrum and the autocorrelation function.

For the pulse, all the Fourier components comprising it are in phase at one instant, when the pulse occurs. With the noise, however, although the same frequency components are involved, they have random phase distribution, and therefore produce a random signal.

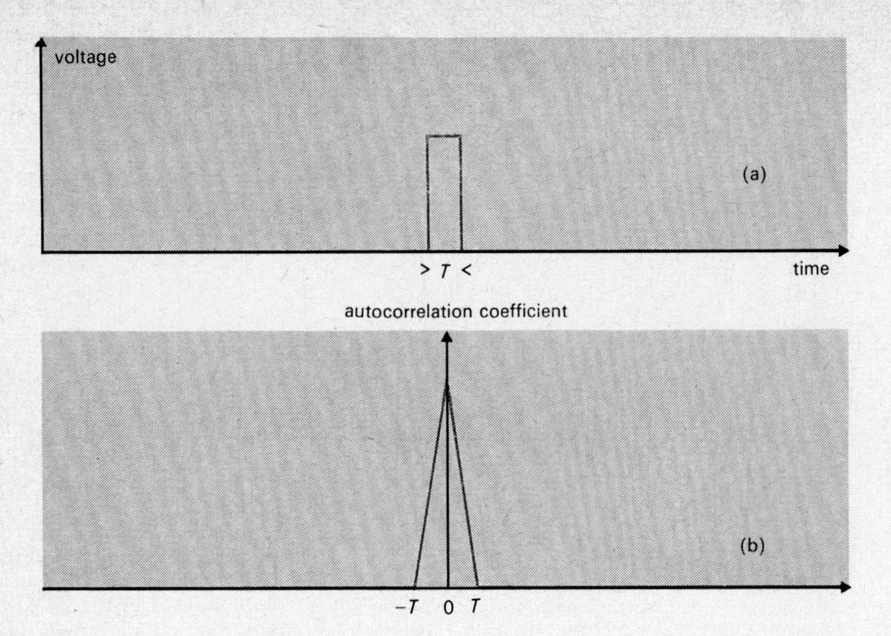

Figure 27 (a) A rectangular pulse and (b) its autocorrelation function

Example 3 Pseudo-random noise

A particularly important type of signal, which we shall encounter again in Unit 13, is called a *chain code*, or a *pseudo-random binary sequence*. It is actually a periodic signal, but within each period it has a number of random noise-like properties, so it is often called pseudo-random noise. Some of its properties were described in the *Technology Foundation Course** Unit 3 'Speech, Communication and Coding' and some will be described in Unit 13 of this course, but for the moment its main interest lies in the fact that it is a binary sequence (i.e. it has only two values) and is therefore a simple one on which to perform autocorrelation calculations.

chain code, pseudo-random binary sequence

A short version of this type of signal, with binary values ± 1 is given below:

$$-1\ -1\ +1\ +1\ +1\ -1\ +1\ +1\ -1\ -1\ +1\ -1\ +1\ -1\ -1.$$

One period of this particular signal is shown. It is only fifteen samples long. That is, the signal repeats every fifteen samples.

Table 3 shows the calculation of the autocorrelation function. The first row represents the signal. The second row shows the same signal shifted two places to the right, so that it is properly placed for calculating the third autocorrelation coefficient. The products are shown, for each sample, within the period. The average of these values is $-1/15$.

Table 3 Calculation of the autocorrelation function for a pseudo-random binary sequence

			←						period of signal								→		
waveform	−1	−1	−1	−1	+1	+1	+1	−1	+1	+1	−1	−1	+1	−1	+1	−1	−1	−1	−1
2 shifts			−1	−1	−1	−1	+1	+1	+1	−1	+1	+1	−1	−1	+1	−1	+1	−1	−1
products			+1	+1	−1	−1	+1	−1	+1	−1	−1	−1	−1	+1	+1	+1	−1		

**The Open University (1972) T100* The Man-made World: Technology Foundation Course, *Open University Press.*

SAQ 9

SAQ 9

Cut out a piece of paper, mark the binary sequence on it and place it next to the signal shown in Table 3 at various values of parametric time. For each value work out the autocorrelation coefficient. Take values of parametric time of −2, −1, 0, +1, +2 and +3 sample intervals. (You already have the value for +2 sample intervals, so you have five other values to work out.)

You should have found that for all values of parametric time, except zero, the autocorrelation coefficient was −1/15, while for zero it was +1. Thus the autocorrelation function appears as in Figure 28(b) for this sequence, which is shown as a binary waveform in Figure 28(a).

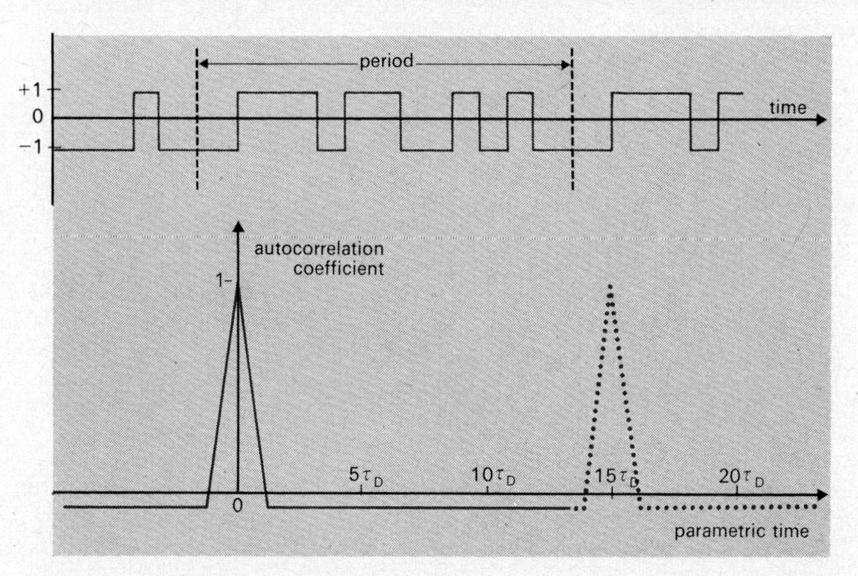

Figure 28 (a) A fifteen-digit pseudo-random binary sequence; (b) its autocorrelation function. Since the binary sequence repeats every fifteen digits, the autocorrelation function is periodic

Because the pseudo-random binary sequence is periodic, the autocorrelation function repeats with the periodicity of the original signal. (A shift of fifteen intervals of parametric time is the same as no shift at all.) The point to note is that like random noise the main feature of the autocorrelation function of pseudo-random noise is a 'spike' but that in this case it occurs at periodic intervals in parametric time. The periodic autocorrelation function is a result of the periodic nature of the pseudo-random binary sequence.

If the method by which an autocorrelation function of any signal can be calculated from its samples is clear, we can now consider the purposes of carrying out this calculation.

One of the reasons for finding the autocorrelation function of a signal is that it can be used to obtain the power density spectrum of the signal.

This is because the power density spectrum $G(f)$ is related to the autocorrelation function $R(\tau)$ via the *Wiener–Khintchine relationship*

Wiener–Khintchine relationship

$$G(f)=2\int_{-\infty}^{\infty} R(\tau)\cos(2\pi f\tau)\,\mathrm{d}\tau$$

and because the autocorrelation function is symmetrical this can be written as

$$G(f)=4\int_{0}^{\infty} R(\tau)\cos(2\pi f\tau)\,\mathrm{d}\tau.$$

This way of finding the power density spectrum of a signal is used for continuous random signals.

In practice, for aperiodic signals, the autocorrelation coefficients approach closer and closer to zero for values further and further from $\tau=0$. This means that the integration need not be performed to infinity and that the power density spectrum of a random signal can be found readily from its autocorrelation function.

You are not expected to remember the form of the Wiener–Khintchine relationship, but you are expected to remember that it enables the power density spectrum to be calculated from the autocorrelation function.

Because the power density spectrum can be obtained from the autocorrelation function, the autocorrelation function must contain all the information present in the power density spectrum. Thus it is sometimes possible to deduce information about the power density spectrum simply by examining the autocorrelation function. In these cases the subsequent evaluation of the Wiener–Khintchine relationship can be avoided.

To see what sort of things can be deduced from the autocorrelation function, let us consider some examples of the relationship between power density spectrum and autocorrelation function. Figure 29(a) shows the autocorrelation function of a signal which has a uniform power density spectrum between 0 Hz and B. Notice that the width of the main spike in the autocorrelation function depends upon the bandwidth of the original signal. Figure 29(b) shows the autocorrelation function of a signal with a power density spectrum which is uniform up to a higher frequency. Notice that the central spike in the autocorrelation function is narrower. So these diagrams illustrate one of the things that we can say about the power density spectrum of a signal from an examination of the autocorrelation function of the signal.

The narrower the central maximum ('spike') of the autocorrelation function is, the higher are the frequencies present in the signal.

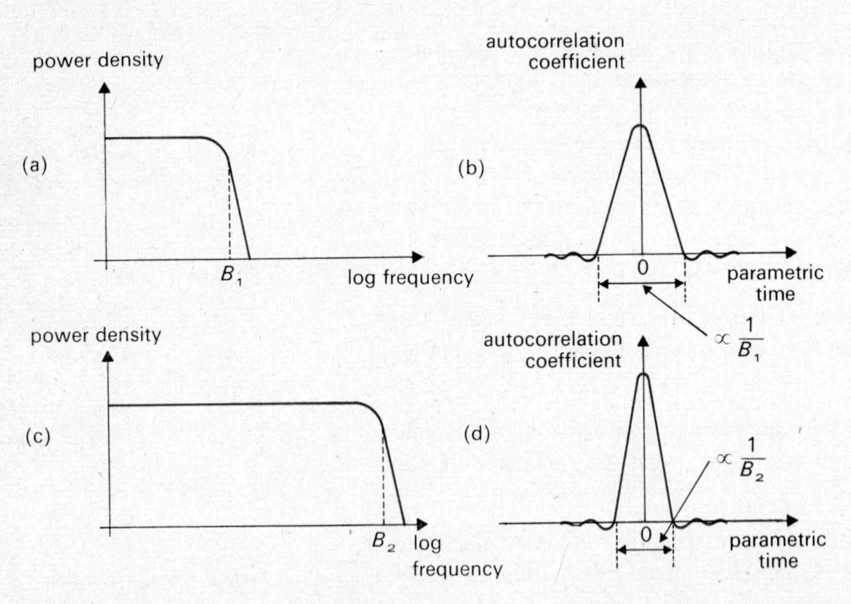

Figure 29 (a) A uniform power density spectrum for a signal and (b) its autocorrelation function. (c) A uniform power density spectrum for a signal with a wider bandwidth and (d) its autocorrelation function. The autocorrelation function for the signal with the wider bandwidth has a sharper 'spike'

But we cannot deduce anything about the shape of the original signal, since neither the power density spectrum nor the autocorrelation function contain any phase information.

One other thing we can deduce from the autocorrelation function of a signal is whether the signal is periodic or not. We have seen already that periodic signals such as a sinusoid, a square wave and a pseudo-random binary sequence have periodic autocorrelation functions. In fact *all periodic signals have autocorrelation functions which are periodic in parametric time.* Furthermore, *the period of the autocorrelation function in parametric time is the same as the period of the signal in real time.*

Finally, *one can deduce the shape of the power density spectrum from the shape of the autocorrelation function.*

To help you to do this I have drawn some autocorrelation functions and their corresponding power density spectra in Figure 30. The spectrum shown in Figure 30(a) is not strictly realizable but it illustrates the limiting case of a band-limited signal with an infinitely sharp cut-off.

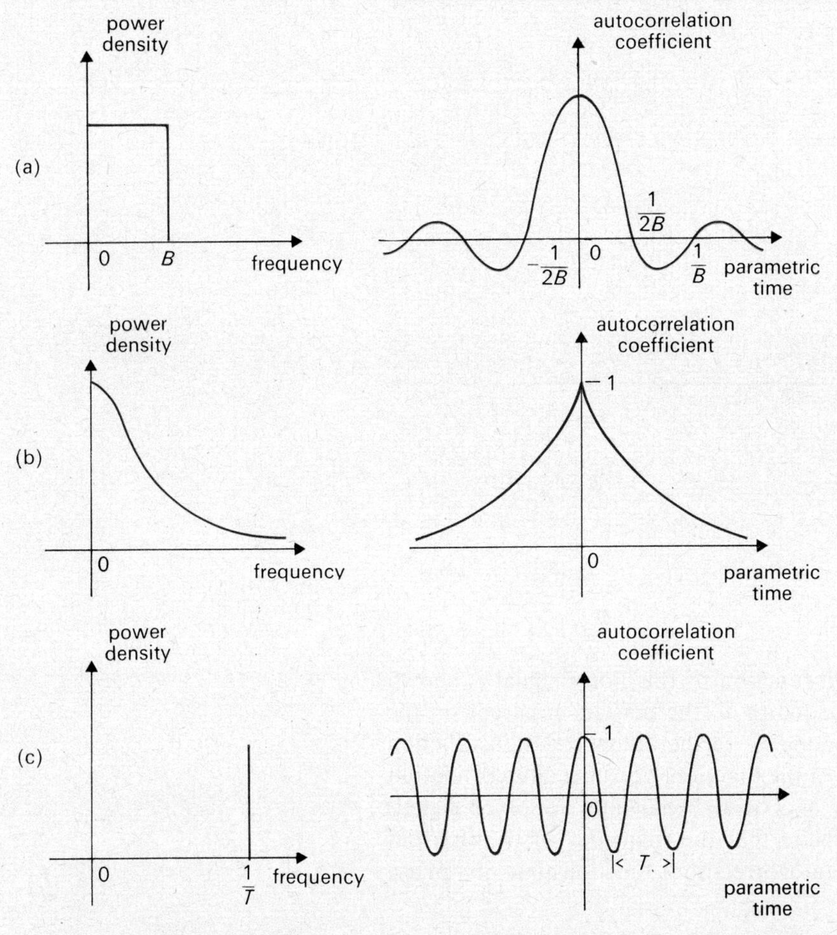

Figure 30 Some power density spectra and their corresponding autocorrelation functions

Now, it is possible to use a knowledge of the properties of the autocorrelation function to detect the presence of a periodic signal in noise.

4.1 The autocorrelation function as a means of detecting signals in the presence of noise

To see how autocorrelation makes this possible, let us consider an example.

Suppose a sinusoidal signal (Figure 31a) is affected by the addition of white noise (Figure 31b). A portion of the resulting signal is shown in

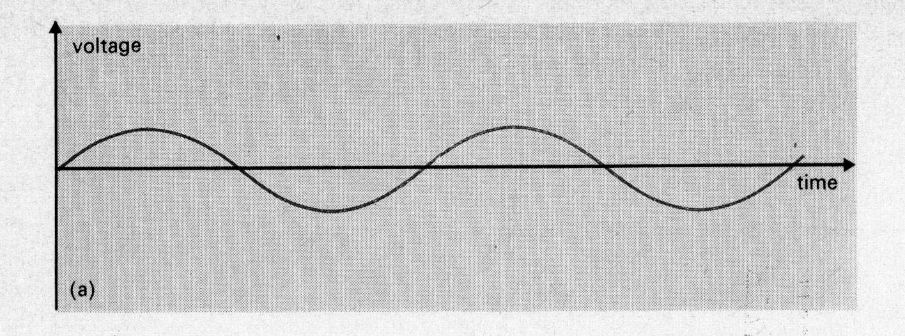

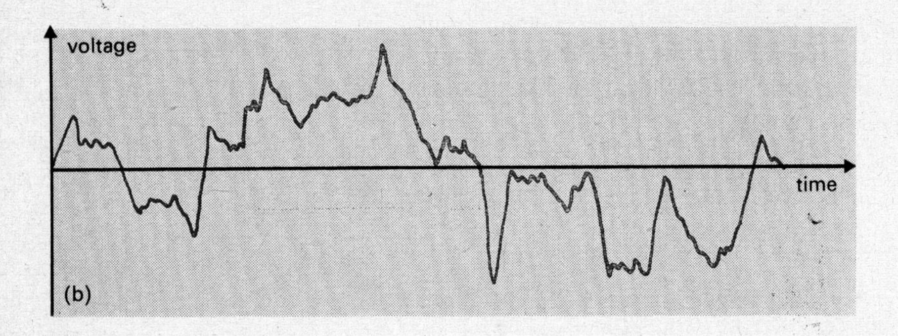

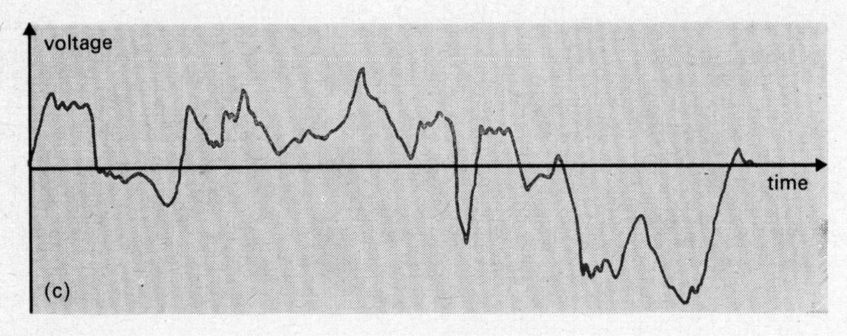

Figure 31 *(a) A sinusoid; (b) a segment of Gaussian noise; (c) the result of adding the noise to the sinusoid*

Figure 31(c). The autocorrelation function of this noisy signal is shown in Figure 32.* The 'spike' corresponding to the noise is apparent on the y-axis, while the sinusoid corresponding to the autocorrelation function of a sine wave dominates the rest of the function. Thus the autocorrelation function reveals that it consists of a wide-bandwidth noise (or signal) and a sinusoid added together. Note that the frequency of the sinusoid can be determined from the autocorrelation function but its phase can not.

> How would you calculate the frequency of the sinusoid from an autocorrelation function such as that shown in Figure 32?
>
> ---
>
> The frequency of the sinusoid is 1/(the period of the sinusoid in the autocorrelation function).
>
> ---

The longer the observed segment used for the autocorrelation process, the easier it is to separate the two parts of the autocorrelation function. If the sinusoid does not exist for a sufficient number of cycles in the segment, it may be impossible to distinguish the two parts of the autocorrelation function at all. The length of the signal which needs to be used to separate the

**This autocorrelation function can be thought of as the result of using a longer and longer segment of the signal to plot an autocorrelation function until a further lengthening of the segment produces no significant change in the autocorrelation function obtained.*

two parts of the function depends upon the relative sizes of the signal and noise and upon the frequencies present in the signal and the noise.

Figure 33 shows the autocorrelation function of another noisy signal. Is it possible to distinguish the parts of this function and conclude what the components of the original signal must have been?

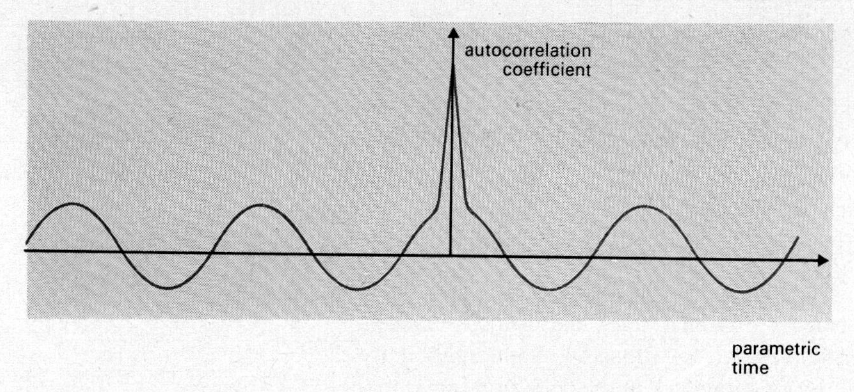

Figure 32 The autocorrelation function of the noisy sinusoid drawn in Figure 31(c)

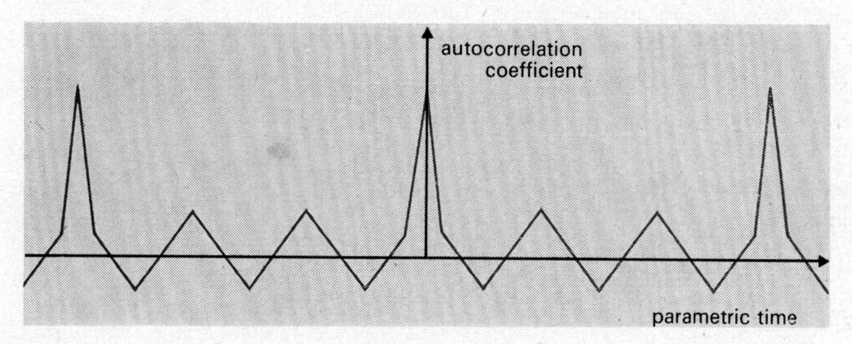

Figure 33 Autocorrelation function of a signal with added noise

We can separate the autocorrelation function into two parts: a periodic 'spike' autocorrelation function superimposed on a periodic triangular-shaped function. But, because the autocorrelation function does not describe a waveform completely, it is not possible to say what the waveforms that caused these separate autocorrelation functions looked like.

Now, suppose I had put the problem a slightly different way. Suppose we knew that the only possible waveforms that could be transmitted to us were a square wave or simply a constant zero voltage level and that the interfering noise was pseudo-random noise. Then the triangular part of the autocorrelation function of Figure 33 would tell us that the square wave was being received during the observation interval used to obtain this autocorrelation function. We could also determine the frequency of the square wave being received by finding the frequency of the triangular autocorrelation function. If the square wave were not present, the autocorrelation function would consist of just the periodic impulsive part of Figure 33.

As we can see from the examples, the autocorrelation function provides a means of detecting signals affected by noise by exploiting the differences in the shape of the autocorrelation functions of the wanted signal and the interfering noise.

A better, but slightly more complicated, way of detecting the presence of signals is to use cross-correlation. I shall be describing this technique in Unit 13.

4.2 Summary of Section 4

The process of autocorrelation establishes the relationship between a signal and a time-shifted version of itself.

To find the autocorrelation function of a waveform:

1 Lay the waveform and its duplicate side by side but shifted in time relative to each other by a time delay known as the *parametric time*.

2 For each chosen value of parametric time, obtain the autocorrelation coefficient, that is,

(a) obtain the product of each sampled value of the waveform and the sampled value of its delayed duplicate, which are presented to the multiplier at the same instant;

(b) take the average of all these products.

3 Plot the values of the autocorrelation coefficients against the corresponding values of parametric time.

The power density spectrum can be obtained from the autocorrelation function by using the Wiener–Khintchine relationship. Sometimes it is not necessary to do this because, by examining an autocorrelation function, several things can be deduced about the spectrum of the signal.

1 The narrower the central maximum of the autocorrelation function, the higher are the frequencies present in the signal.

2 If an autocorrelation function is periodic, then the signal which produced it is also periodic and has the same period in real time as the autocorrelation function has in parametric time.

3 The shape of the power density spectrum can be deduced from the shape of the autocorrelation function.

Finally I have shown that it is possible to use a knowledge of the properties of the autocorrelation function to detect the presence of a periodic signal in noise.

The autocorrelation function completes the set of parameters I introduced in Unit 11. The mean, r.m.s., standard deviation and probability density function are concerned with characterizing the waveform in terms of the instantaneous values of the signal. The power density spectrum and autocorrelation function are concerned with characterizing the frequency content of a signal. The methods of obtaining these parameters, some of their uses and the relationships between them are summarized in Table 4. You should spend some time studying Table 4 as it summarizes many of the ideas introduced in the last one and a half units.

For the remainder of the unit I shall be describing ways of changing the frequency content of a signal so as to make it more suitable for transmission over a noisy channel.

Table 4 Summary of the signal parameters introduced in Units 11 and 12. Included are some of their uses

Parameters describing the values of the signal

Mean

$$\bar{x} = \frac{\sum_{i=1}^{n} x_i}{n}$$

Use

Diminishing the fluctuations in a noisy signal to reveal the constant noise-free value.

Standard deviation

$$\sigma=\sqrt{\frac{\sum_{i=1}^{n}(x_i-\bar{x})^2}{n}}$$

Use

Characterizing the dispersion of a signal.

Probability density function

Can be estimated from a histogram obtained from a waveform or from a knowledge of the causes of the signal.

Use

Estimating likely error rates.

Parameters describing the frequency content of the signal

Line spectrum

Can be obtained for a finite segment of a signal by finding the Fourier series which describes the corresponding periodic signal.

Use

Characterizing amplitude and phase of the signal frequency components of many deterministic waveforms.

Power density spectrum

Can be obtained from line spectra of segments of a signal or by using a spectrum analyser or from the autocorrelation function

Use

Characterizing the frequency components of a random signal.

↑ *Wiener–Khintchine relationship*

Autocorrelation function

Can be obtained using a correlator.

Use

Characterizing the frequency content of signals. Can be used to detect periodic signals in noise.

The transmission of signals through noisy channels

So far I have talked about the measurement of signals in the presence of noise by averaging, filtering and autocorrelation. This noise may be unavoidably introduced when the signal is being transmitted from its origin (maybe a transducer) to a receiver. That is, the noise is introduced on the transmission channel. Such a channel is termed a *noisy channel.*

noisy channel

Let us briefly consider two examples of a noisy channel in an instrumentation system.

If the instrumentation system is measuring a strain on the shell of an aircraft, then it is likely that the communication channels may be long lengths of electrical cable, each one of which will be picking up a certain amount of electrical noise.

Another instrumentation system may involve the transmission of information about the physical condition of astronauts. In this case the communication channel will be a radio channel from the spacecraft to Earth. Any receiver of this signal is bound to pick up interstellar noise as well as the signal transmitted by the spacecraft. In both these cases the amount of noise picked up from the channel may be sufficient to degrade the message being transmitted to a significant extent. The amount of degradation that is regarded as significant depends on the use to be made of the message signal at the receiver.

In fact there are many ways of altering a signal to make it suitable for transmission over a noisy channel. In this section I shall describe some of them and the topic will be continued in Unit 13.

Now, the simplest way to overcome noise picked up during transmission is to increase the power of the signal at the source (e.g. to shout in a noisy factory to make yourself heard). But it is not always possible to amplify the signal sufficiently to render the interference negligible, because the power required may be more than you have available. Under these circumstances it is worth studying how the signal can be modified to overcome the effects of noise, without having to increase its maximum power.

In the case of the signal transmitted from the spacecraft, it is necessary in any case to modify the output of the transducer by a process of 'modulation' in order to transmit it by radio waves. However, even for the case of channels consisting of wires it is often worth introducing a system of modulation simply for the purposes of combating noise.

Now what is modulation? *Modulation* is the modification of a *carrier* waveform* in response to the information carried in a signal (e.g. the output of a transducer). The process of recovering the original signal from the modulated signal is called *demodulation.*

modulation carrier

demodulation

The most familiar use of modulation is in radio transmission. The radio signals transmitted by the BBC on the medium and long wave bands are 'amplitude modulated'. An example of *amplitude modulation* (AM) is shown in Figure 34.

amplitude modulation

Figure 34(a) shows a sawtooth waveform, which we shall regard as the wanted signal, and Figure 34(b) shows a sinusoidal carrier whose frequency is much higher than the fundamental frequency of the sawtooth.

The amplitude-modulated waveform is shown in Figure 34(c). In amplitude modulation the instantaneous amplitude of the carrier is

**This is often a sinusoid but it may be another type of waveform.*

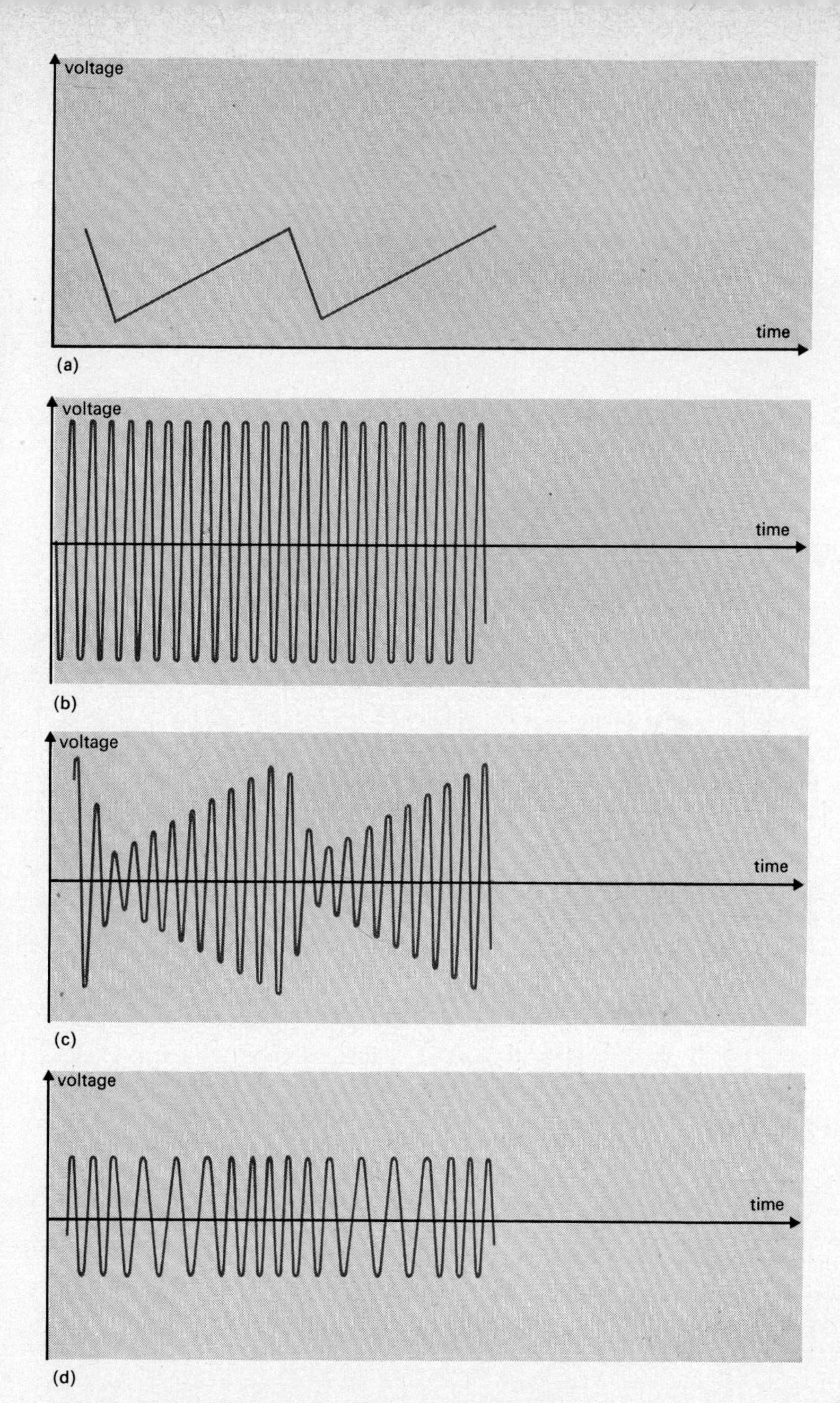

Figure 34 Amplitude and frequency modulation: (a) the message signal; (b) a sinusoidal carrier; (c) amplitude modulation; (d) frequency modulation

determined by the instantaneous value of the modulating signal. As the signal goes more positive, the amplitude of the sinusoidal carrier increases, and vice versa.

frequency modulation

Figure 34(d) shows a different kind of modulation. This is *frequency modulation* (FM), and, as you can see, the amplitude of the modulated waveform does not vary. In this case the *frequency* of the carrier varies in response to the instantaneous value of the message signal. The more positive the message signal becomes, the higher the frequency of the modulated carrier.

pulse code modulation

Figure 35 shows an entirely different kind of modulation called *pulse code modulation.* (This is often abbreviated to PCM.) In pulse code modulation the message signal is first sampled to obtain a series of

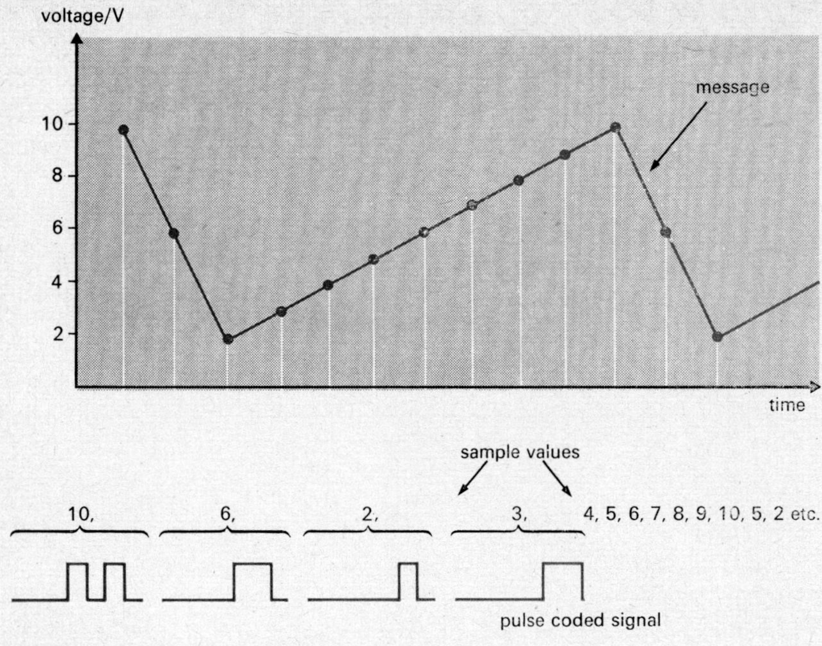

Figure 35 An example of pulse code modulation

sample values which completely describe the signal. This is the sampling process described in Unit 11. These sample values are then converted into groups of binary pulses. Each group represents one sample value as a binary number.

For example, suppose that there are seven binary digits in each group. (There can be any number in a group, the more digits there are, the more accurately the sample values can be represented by binary numbers.) Suppose also that the sample values to be represented are in the range from 0 to 127 mV with quantizing intervals of 1 mV. Thus the code for 127 mV is

1 1 1 1 1 1 1,

and the seven-digit binary code for 10 mV is

0 0 0 1 0 1 0.

The binary pulses that represent this value are shown in Figure 36. Here the positive voltage indicates the binary digit 1 and the negative voltage the binary digit 0. It is this waveform of Figure 36 that is transmitted along the channel. The process continues with the next sample value being converted to a binary waveform and so on.

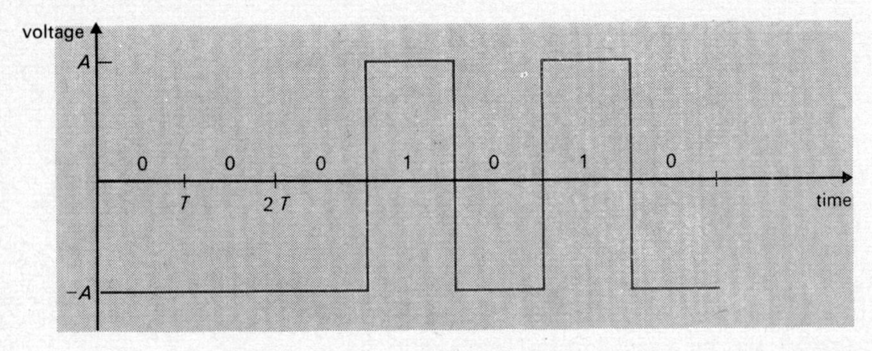

Figure 36 The PCM binary waveform representing 10 mV

These are the three main types of modulation in common use, though there are many variants of all three. We shall be considering each type briefly in the next few sections. Before doing so we must note the possibility of achieving even better noise immunity in the signal by compensating for the effects of noise with the application of error-correcting codes. Error-correcting codes are most successfully applied in relation to some form of pulse code modulation, as we shall see in Unit 13.

It should be appreciated at the outset too that combinations of modulation methods are possible. For example, as I have just mentioned, pulse code modulation is a conversion of an analogue signal into a succession of two-level binary pulses. This two-level binary signal could then be regarded as the message signal with which to frequency-modulate a carrier. Under these circumstances two different carrier frequencies correspond to the two binary values of the PCM signal. This is sometimes called frequency shift keying, but the point to note here is that the message signal has undergone two modulation processes. It has first been sampled and converted into a PCM signal and then this PCM signal has been used to modulate a carrier.

In the next few sections I want to explain how it is that these different types of signal have different properties in relation to interference. In order to make satisfactory comparisons between the various methods, I shall assume that there is a given maximum power available from the source in each case. For example, if we consider the case of a satellite in space, the power is ultimately limited by the amount of power available from the solar cells being used to convert sunlight into electric power. Thus, whatever modulation method is used, the maximum available power is limited. But as far as we are concerned here in comparing one kind of modulation with another, we shall assume that all methods are equally limited as to maximum power in the same way.

Let us also assume that the characteristics of the interference – the channel noise – are the same in each case. That is to say, the noise has the same statistical properties and the same power density spectrum. For simplicity, let us assume the channel noise to be white noise, that is, noise for which the power density spectrum is uniform over the bandwidth of interest.

5.1 Amplitude modulation

To simplify the description of amplitude modulation (AM), let us consider a sinusoidal 'message signal' modulating a sinusoidal carrier of higher frequency. This is illustrated in Figure 37. First let us consider the bandwidth of the modulated waveform shown in Figure 37(b). Since this is not a sinusoid it follows that there must be more than one frequency present. The question is, what frequencies are present and within what bandwidth do they lie?

Suppose the carrier is described by the equation

$$v = K \sin \omega_c t.$$

Here K is the amplitude and ω_c is the angular frequency of the carrier.

Now, in order to express the modulation of the amplitude of the carrier, we must replace the amplitude K by a term which varies with the modulating signal. From Figure 37(b) we can see that we want an expression which has the value A when the instantaneous value of the message signal is zero, but which rises above A when the message signal has a positive value and falls below A when the message signal has a negative value. For the sinusoidal message signal of Figure 37(a), the expression for K is

$$K = A(1 + m \cos \omega_m t).$$

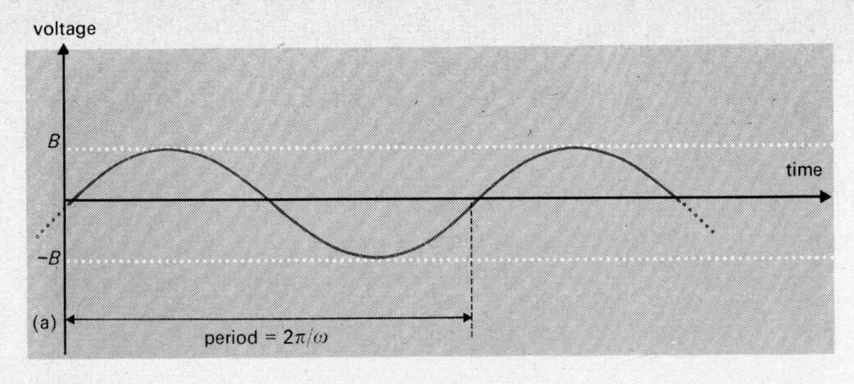

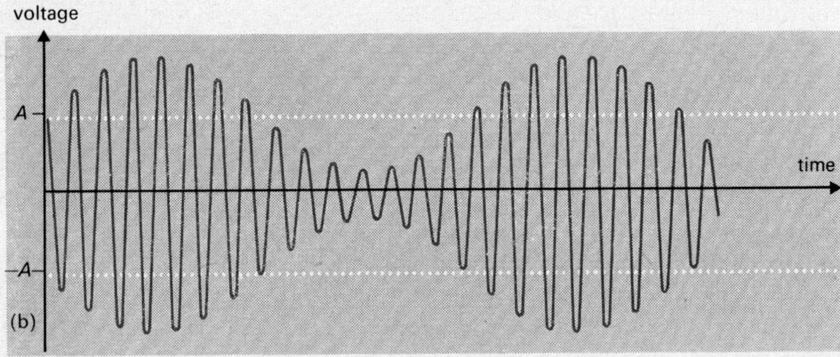

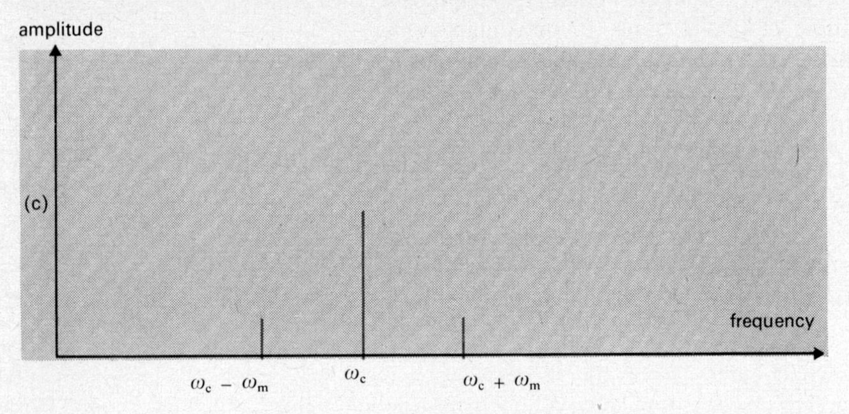

Figure 37 Amplitude modulation: (a) a sinusoidal 'message signal' of angular frequency ω_m; (b) the amplitude-modulated carrier of angular frequency ω_c; (c) the spectrum

Here m is the amplitude of the message signal relative to the amplitude A of the carrier. The actual amplitude of the message signal is Am. The frequency ω_m is the angular frequency of the message signal.

Check with Figure 37(b) to confirm this.

The equation for the final modulated signal is found by substituting for K in the original equation for the carrier,

$$v = A(1 + m \cos \omega_m t) \sin \omega_c t.$$

This equation can be manipulated using standard trigonometrical relationships,* and can be reduced to the form

$$v = A[\sin \omega_c t + \tfrac{1}{2}m \sin(\omega_c + \omega_m)t + \tfrac{1}{2}m \sin(\omega_c - \omega_m)t]. \quad (1)$$

Check this manipulation using 'Mathematics for Instrumentation' if necessary.

Thus, as you can see, the waveform can be broken down into three frequency components, one at the carrier frequency ω_c, one at a

* $\sin(A+B) = \sin A \cos B + \cos A \sin B,$
$\sin(A-B) = \sin A \cos B - \cos A \sin B,$
adding: $\sin(A+B) + \sin(A-B) = 2 \sin A \cos B.$

frequency corresponding to the sum of the carrier and message frequencies $\omega_c+\omega_m$ and one corresponding to the difference between the carrier and message frequencies $\omega_c-\omega_m$. See Figure 37(c).

So the overall bandwidth of the modulated signal is the difference between the highest and lowest frequencies, namely

$$(\omega_c+\omega_m)-(\omega_c-\omega_m)=2\omega_m.$$

What are the amplitudes of the three components of the amplitude-modulated signal?

From equation (1) you can see that the carrier amplitude is A and the sinusoids to each side of it have an amplitude $\frac{1}{2}Am$.

This modulation process has thus not only produced new, 'side', components on each side of the carrier, it has also transformed the frequency of the message from ω_m to around ω_c.

If $f_c=\omega_c/(2\pi)=1$ MHz and $f_m=\omega_m/(2\pi)=1$k Hz, what frequencies are contained in the complete amplitude-modulated signal?

1.001 MHz, 1.000 MHz, 0.999 MHz.

Any message signal can be represented as a sum of sinusoidal frequency components, as was shown earlier in the unit. Each component modulates the carrier as I have just explained, resulting in a band of 'side' signals each side of the carrier frequency. These are called *sidebands*. With amplitude modulation there are two, one above and one below the carrier, containing the sum and difference frequencies of the carrier frequency and the frequency components of the message signal. Thus, as illustrated in Figure 38, the bandwidth of an amplitude-modulated signal is just twice the bandwidth of the message signal itself.

sideband

We must remember this result when we come to consider the amount of noise power interfering with the signal.

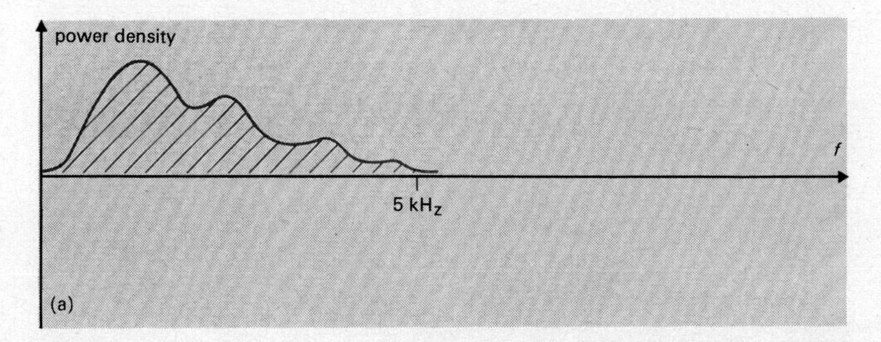

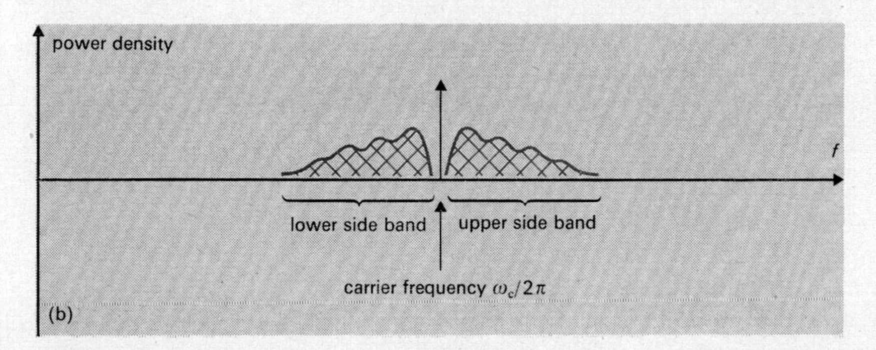

Figure 38 The spectrum of an amplitude-modulated signal: (a) the spectrum of a message signal such as a segment of speech; (b) the spectrum of the corresponding amplitude-modulated signal. (*Here, to scale, the carrier signal is about* 20 kHz, *since each sideband occupies* 5 kHz.) *The sidebands occupy frequency ranges which are the sum and difference of the carrier and the message frequencies*

The maximum permitted peak output voltage of the transmitting device (e.g. a transistor) determines the maximum peak voltage of the modulated signal. It is not possible to increase the message (modulating) signal amplitude indefinitely because the larger values of the modulated signal would be clipped or distorted by the limitations of the transmitting device. Thus the idea of increasing the amplitude of the message signal in order to overcome the effects of noise is limited by the allowable peak voltage (or power) of the transmitter. We shall find that this is not the case when frequency modulation is involved.

Before we look at the way in which noise affects an amplitude-modulated signal, we must consider how a noise-free amplitude-modulated signal is demodulated.

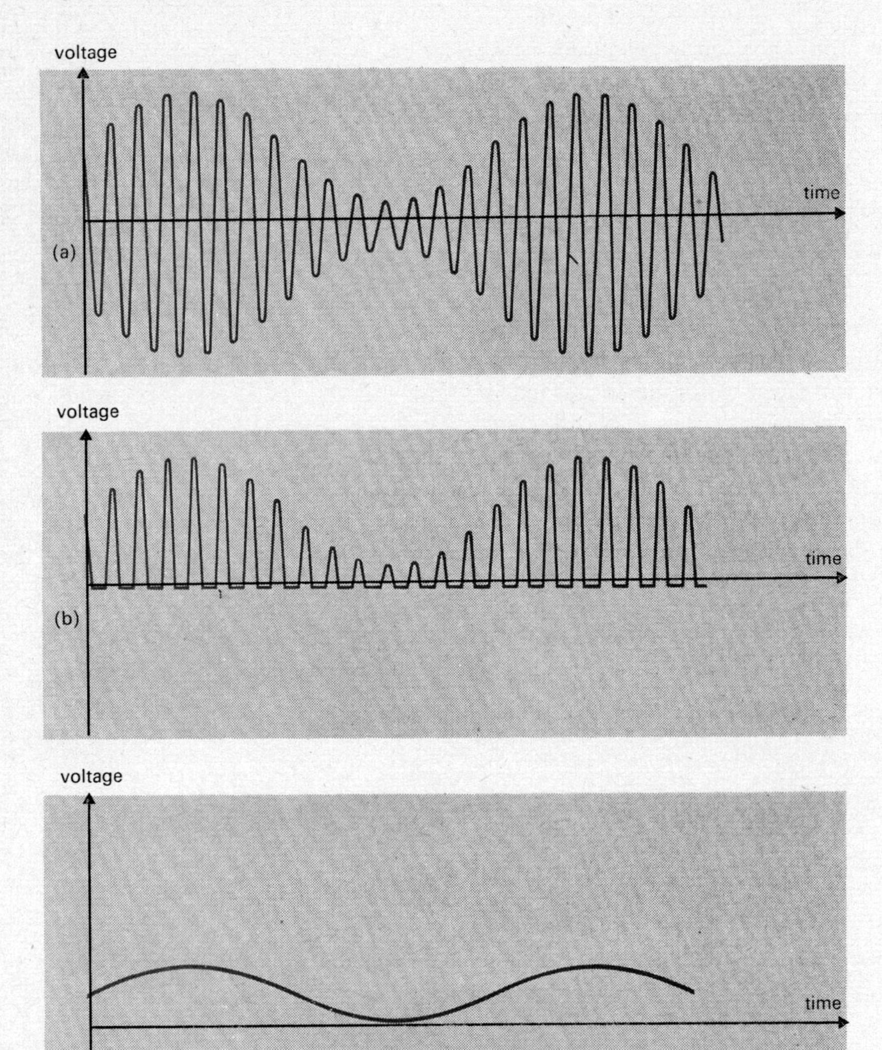

Figure 39 A method of AM demodulation: (a) the amplitude-modulated waveform; (b) the result of rectifying the amplitude-modulated waveform; (c) the result of filtering the waveform shown in (b)

Figure 39(a) shows a carrier amplitude modulated by a sinusoid. The first stage in the demodulation process is to feed this signal into a rectifying circuit. This circuit allows currents caused by the signal voltage to flow in one direction but not in the opposite direction. By connecting the rectifying circuit appropriately the waveform in Figure 39(b) is obtained. This waveform now consists only of the positive parts of the original signal. Notice that the envelope of this new waveform has the same shape as the original modulating sinusoid. The waveform of Figure 39(b) is very similar to a series of sample pulses of the modulating signal. The way in which the modulating signal is recovered is to apply this rectified waveform to a low-pass filter. The filter removes the high-frequency components due to the pulses and allows the low-frequency modulating signal component to pass through unaffected. The output of such a filter when the rectified signal is applied to it is shown in Figure 39(c).

A simplified schematic diagram of an AM demodulator is shown in Figure 40. The input filter is a band-pass filter which passes just the band of frequencies present in the required amplitude-modulated signal. This filter reduces the interference from signals outside its pass band.

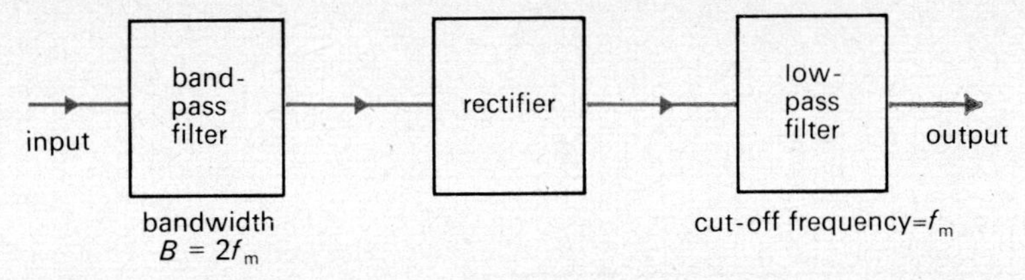

Figure 40 A simple AM demodulation system for a carrier amplitude modulated by a signal whose spectrum extends from 0 Hz *to* f_m

Now, suppose the modulated signal passes through a noisy channel. It will reach the receiver in a form somewhat like that shown in Figure 41(a). After the signal has been demodulated it will appear as in Figure 41(b). In other words, the amplitude of the noise signal will simply be added to the amplitude of the message signal and the degree of interference will depend solely upon the relative amplitudes of the two signals.

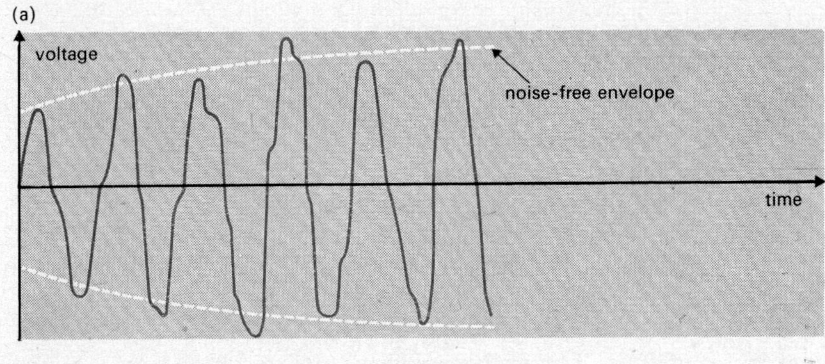

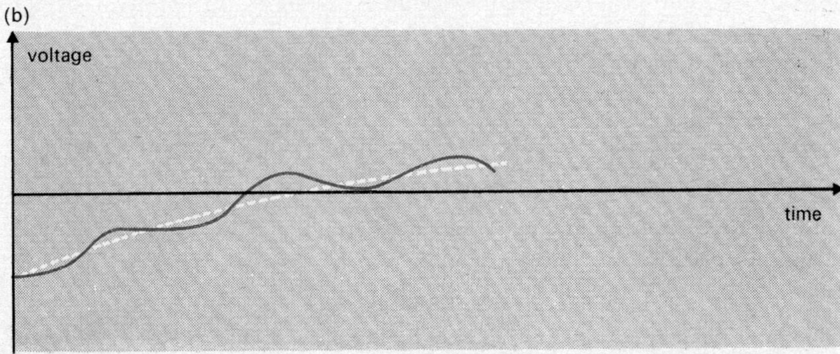

Figure 41 (a) A segment of an AM signal corrupted by noise; (b) the resulting demodulated signal. The original noise-free signal is shown dashed

5.2 Frequency modulation

In frequency modulation (FM) the instantaneous frequency of a carrier is varied in response to the modulating signal. Figure 42(b) shows a carrier which has been frequency-modulated by a sinusoidal message signal. Notice that when the sinusoid has a maximum value the instantaneous carrier frequency is at a maximum and when the sinusoid

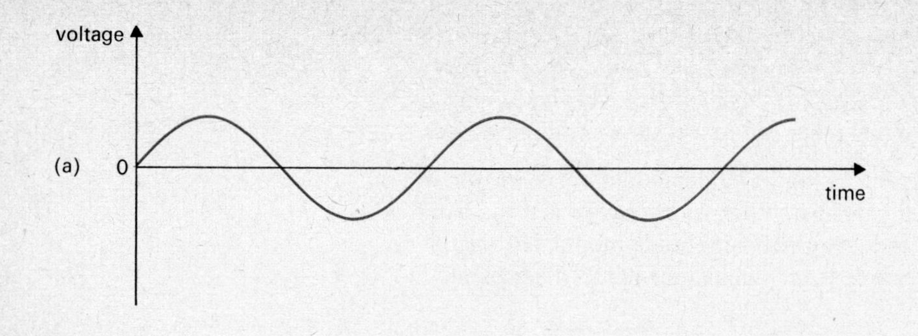

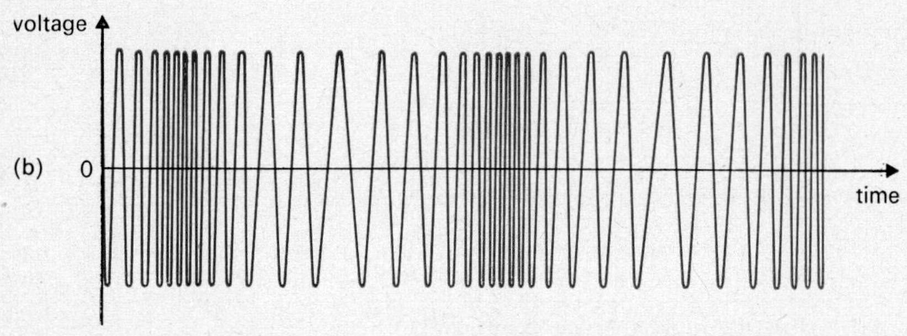

Figure 42 (a) A sinusoidal message signal; (b) a carrier frequency modulated by the message signal

has a minimum value the instantaneous carrier frequency is at a minimum. Also notice that the amplitude ot the carrier does not change.

How is this signal demodulated?

The principle of one scheme is to make this frequency change cause a corresponding change in amplitude and then to use an AM demodulator like the one described earlier.

Figure 43 shows the response of a filter which is constructed so that the frequency of the unmodulated carrier is in the middle of the linear part of the response. When the instantaneous frequency of the carrier increases, corresponding to an increase in the modulating signal, the output of the filter increases in proportion. Similarly, when the instantaneous frequency of the carrier decreases, the output of the filter decreases. Therefore, this filter responds to changes in frequency by producing corresponding changes in amplitude. The final stage in the demodulation is to apply the output of the filter to a rectifying circuit and then a low-pass filter. A schematic diagram of the demodulator is shown in Figure 44.

Now how does noise affect this frequency-modulated signal?

The noise affects both the amplitude and the instantaneous frequency of the frequency-modulated signal. But, because the message is con-

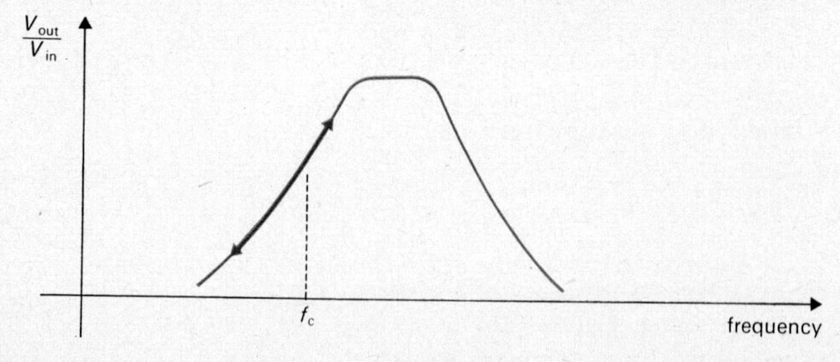

Figure 43 The frequency response of a filter used in an FM demodulator. Marked on the diagram is the carrier frequency f_c and the region over which the instantaneous frequency varies

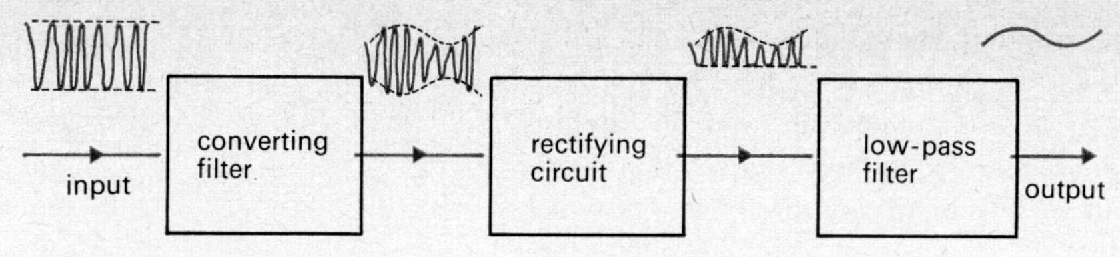

Figure 44 A schematic diagram of a simple FM demodulator, showing waveforms at various points in the circuits for a sinusoidal message signal

tained in the frequency variations, the usual practice is to pass the received frequency-modulated signal through a circuit which limits and removes the amplitude variations before the signal goes to the demodulator. The variations in instantaneous frequency produced by the noise remain and it is these which produce the noise at the output of the demodulator.

Remember that in amplitude modulation, to increase the signal-to-noise ratio at the output of the demodulator, the transmitter power had to be increased. The maximum available transmitter power limited the improvement that could be obtained.

In frequency modulation the signal-to-noise ratio at the output of the demodulator can be increased by increasing the sensitivity of the *modulator*.

Figure 45 shows the output of a frequency modulator for different sensitivities and a sinusoidal message signal. Notice that for increased sensitivity the instantaneous frequency of the carrier changes over a

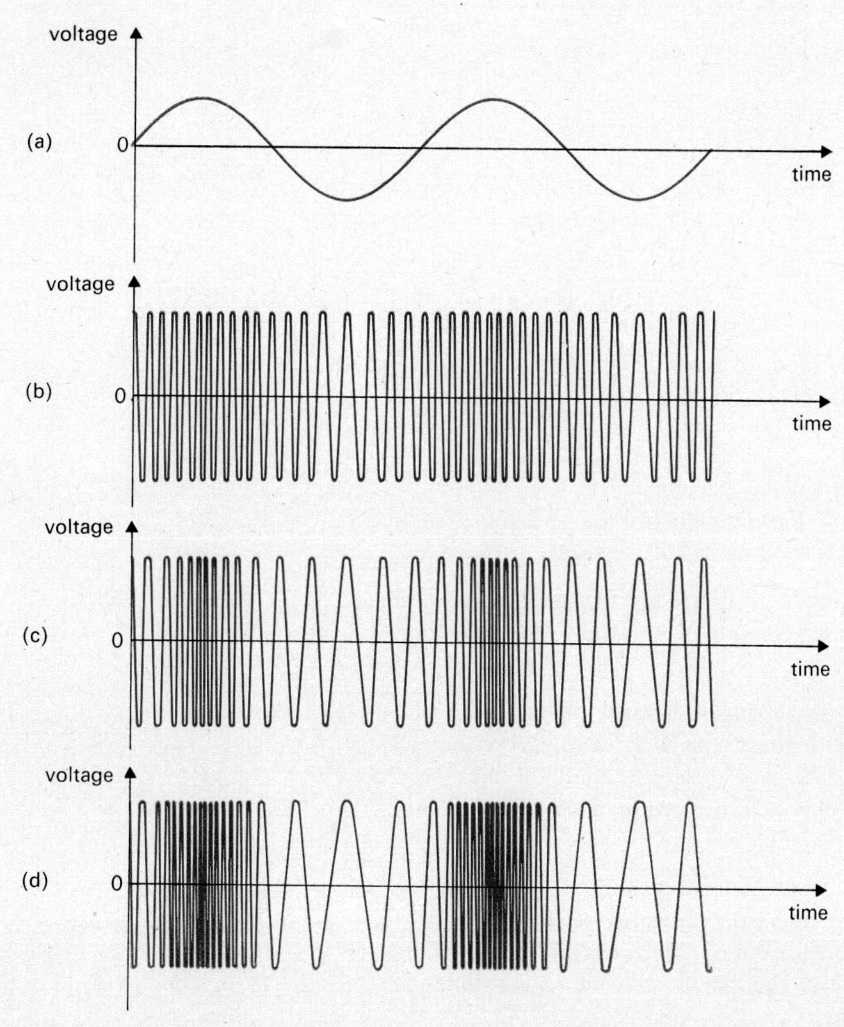

Figure 45 (a) A sinusoidal message signal; (b), (c) and (d) the output of a frequency modulator in response to this message signal for different sensitivities: (b) is the output when the modulator is least sensitive, (d) when it is most sensitive

wider range. This means that at the demodulator the frequency-modulated signal will be converted into an amplitude-modulated signal using a larger portion of the characteristic of the filter, as shown in Figure 46. The larger variations in amplitude will be converted into a larger signal output by the rectifying circuit and low-pass filter circuit.

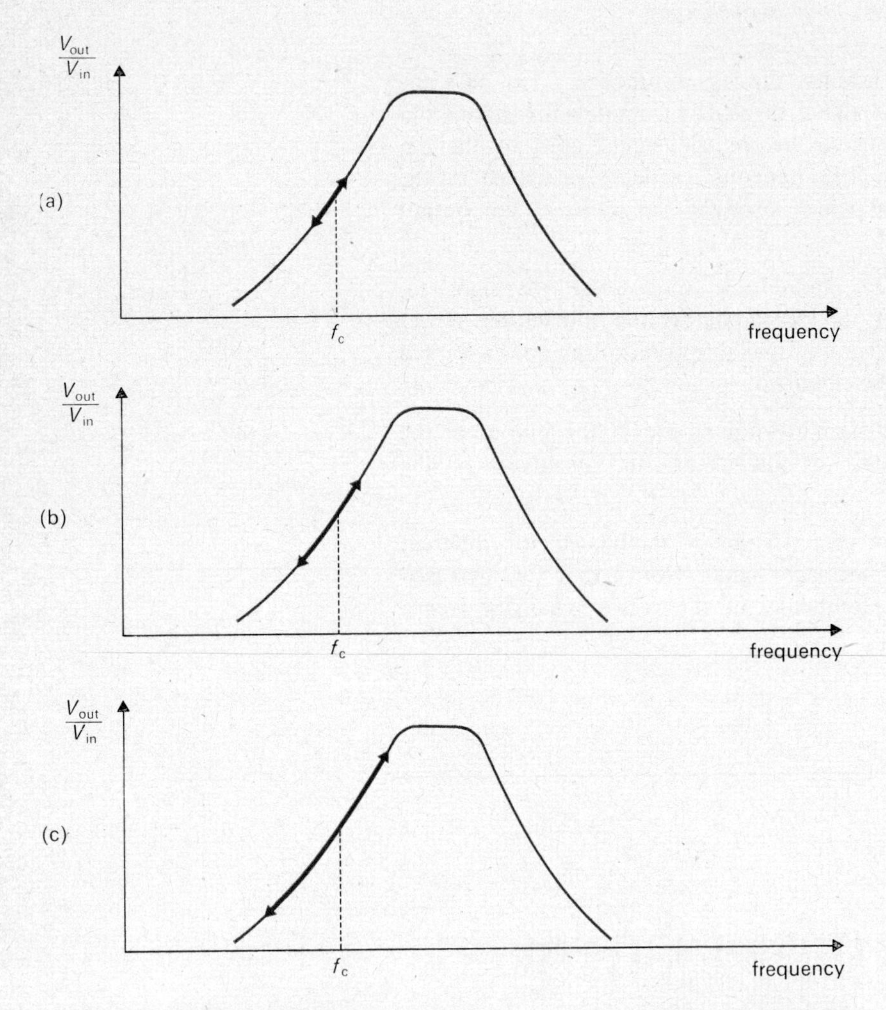

Figure 46 The regions of the converting filter in the demodulator which are used by the three signals shown in Figure 45(b), (c) and (d)

Thus the signal at the output of the demodulator can be increased by increasing the degree of frequency modulation of the transmitted wave.

However, increasing the instantaneous frequency change increases the channel bandwidth that is required to transmit the frequency-modulated signal.

It is easy to understand why this happens because the channel must be able to pass at least the highest and the lowest instantaneous carrier frequencies. As the difference between these instantaneous frequencies increases, the channel bandwidth required for the frequency-modulated signal increases.

Remember that, for an amplitude-modulated signal, the channel bandwidth required was twice the highest frequency of the modulating signal. The frequency components of a *frequency-modulated signal* extend to infinity on either side of the carrier, but their amplitudes fall off rapidly outside a certain range.

Thus the bandwidth that has to be transmitted for a frequency-modulated signal is not the simple value that was obtained for amplitude modulation, because neglecting frequencies outside a certain bandwidth will result in some non-linear distortion being introduced. The amount of distortion which is regarded as being tolerable and, hence, the amount of bandwidth needed for the frequency-modulated signal is very dependent on the particular application.

A very approximate guide to the required bandwidth for a frequency-modulated signal is given by *Carson's rule*. This states that the bandwidth required for a frequency-modulated signal is

Carson's rule

2×(maximum deviation+highest frequency present in message signal).

The *deviation* is the shift in the carrier frequency from its unmodulated frequency. Maximum deviation corresponds to the peaks of the modulating signal.

deviation

This expression is summarized in Figure 47.

As you can see, the significant bandwidth is dependent upon the maximum deviation of the instantaneous frequency from the carrier frequency and upon the highest frequency present in the modulating signal.

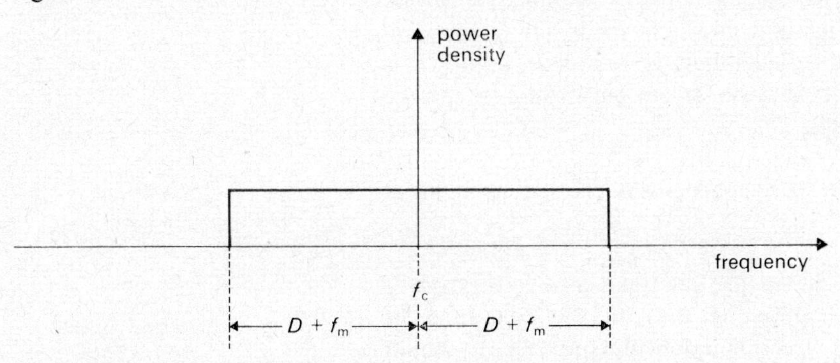

Figure 47 The bandwidth required for a frequency-modulated signal as given by Carson's rule is $2(D+f_m)$, *where D is the maximum deviation and* f_m *is the highest frequency in the message signal*

So, if the deviation is increased in order to increase the output signal from the demodulator, the bandwidth of the receiver must be increased. This then allows more noise to reach the demodulator. However, the message signal output increases to a greater extent than does the noise output and the signal-to-noise ratio is improved.

Frequency modulation, then, enables the improvement of signal-to-noise ratio at the output of the detector without an increase in transmitter power. However, this requires a greater bandwidth than amplitude modulation.

5.3 Pulse code modulation

Considerations which lead to an understanding of the behaviour of pulse code modulation (PCM) in the presence of noise are quite different from those we have discussed in relation to amplitude modulation and frequency modulation. The first point I should make is that the process of converting the sample values into a coded form introduces errors of a special type.

For example, consider the problem of representing a sample value of 9.4 mV using the same seven binary digits as mentioned before, where the binary unit is 1 mV.

The closest binary number to 9.4 is 9, that is, 0001001, and this is the number that would be transmitted. Therefore the modulation itself has introduced an error of 0.4 mV. This is termed a *quantization error*.

quantization error

It is convenient to think of this type of error as equivalent to the addition of a special kind of noise to the message signal. This noise is termed *quantization noise*.

quantization noise

The magnitude of this noise as compared to the magnitude of the signal depends on the number of binary numbers available to describe the sample value. Each available binary number is termed a *quantum level*. (Hence the name of the noise.) For a given signal amplitude range, the larger the number of quantum levels, the smaller is the adjustment which has to be made to each sample value to bring it into line with the chosen set of quantum levels, so the smaller the quantization noise.

quantum level

The bandwidth of the PCM signal depends upon the bandwidth of the original signal and upon the number of binary digits used to represent the sample value.

Suppose the bandwidth of the original signal extends from 0 Hz to f_m, then, by the sampling theorem, the signal must be sampled at a rate of $2f_m$. Now, if the number of quantum levels chosen is 128 (a common number used in practice), then a seven-digit binary code sequence is needed to express this number, because seven binary digits can be arranged in $2^7=128$ different ways.

What is the bandwidth of a signal containing seven digits occurring at a rate of $2f_m$?

The pattern of digits within each sample can take any form from a pattern consisting only of 0s, consisting only of 1s, consisting of alternate 1s and 0s or, indeed, any combination of 1s and 0s. It should be apparent that the kind of pattern containing alternate 1s and 0s will contain the highest-frequency components, as illustrated in Figure 48.

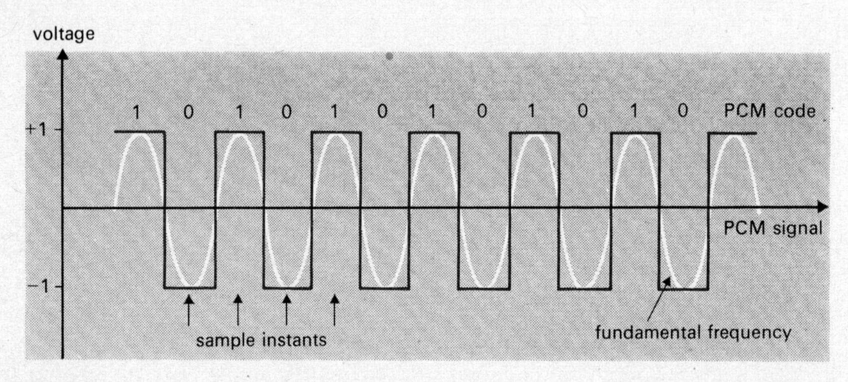

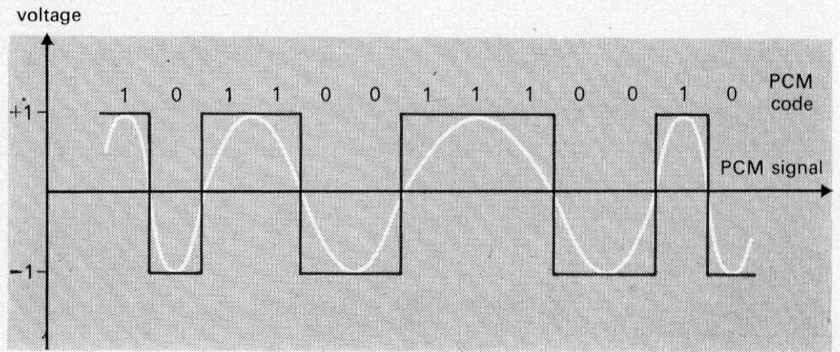

Figure 48 Two PCM signals. The diagram illustrates that the highest frequency occurs in the signal when the coded signal alternates from 1 to 0. Sampling instants of the received signal are indicated

Now, in order to receive this signal without error, all that is necessary for the receiver to do is to distinguish between a 1 and a 0 at an instant midway through the duration of each digit, as indicated by the arrows on Figure 48. If the PCM signal is put through a low-pass filter whose bandwidth is just sufficient to pass a sine wave whose frequency corresponds to the rate at which 1s and 0s alternate (see Figure 48), then the receiver will be able to discriminate correctly between 1s and 0s. Thus, the bandwidth required is half the maximum rate at which the digits are generated by the pulse code modulator. Finally, therefore, we can conclude that the bandwidth of the PCM signal in our example, when 128 quantum levels are used, is $\frac{1}{2}(2f_m \times 7) = 7f_m$.

Thus, by using a seven-digit pulse code modulation and sampling at just twice f_m, we have increased the bandwidth of the message signal by a factor of seven.

The PCM signal is often coded in frequency-modulation terms, in which case there is no difference in voltage level, or power, between a 1 or a 0, just a difference in frequency between the two.

Now, how does noise affect the reception of the PCM signal? The relationship between the amplitude of the noise and the errors produced in the signal is a highly non-linear one. Provided the noise is insufficient to deceive the receiver, then every 1 transmitted will be received as a 1 and every 0 will be received as a 0, even though there is noise in the channel and each pulse is affected by it.

There will only be an error if the received pulses are excessively affected by the noise. If the noise becomes large enough for there to be a significant chance of a random excursion of the noise signal being as large as the amplitude of the PCM signal, then it is possible for the receiver to make a false decision and decide, for instance, that a received digit represents a 1 when it was transmitted as a 0. In fact, there is a kind of threshold. If the noise power is near a certain value, then the number of errors decreases very rapidly as the signal power increases. Conversely, at the critical level a further small increase in noise will produce a rapid increase in the number of errors. The number of errors occurring in a PCM signal is usually quoted as a proportion of the received digits. This proportion is called the *error rate*.

error rate

Figure 49 shows the dependence of error rate upon signal-to-noise ratio for a PCM signal.

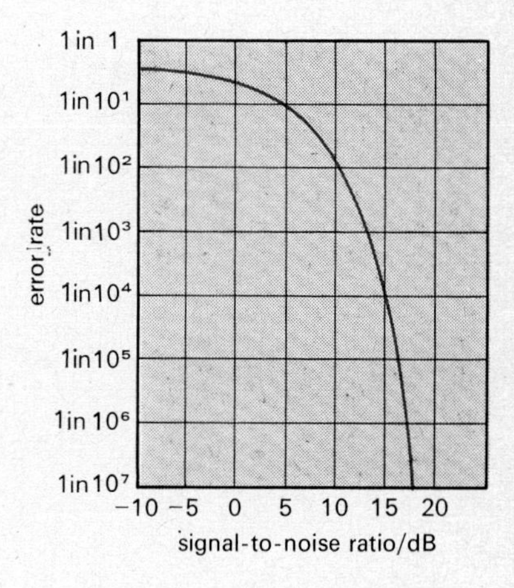

Figure 49 The dependance of PCM error rate on signal-to-noise ratio

Error rate is not the only consideration to be taken into account when assessing the effect of noise upon PCM, because the consequences of an error in one digit may not be the same as the consequences of error in another digit.

For example, in a seven-digit pure binary code the most significant digit has a significance which is sixty-four times that of the least significant digit. (The most significant digit has value $2^6 = 64$, the least significant digit has value $2^0 = 1$.) Thus, if an error in the last digit were to lead to an error of 1 mV, an error in the first digit would lead to an error of 64 mV. Evidently, it is more important to ensure that the most significant digits are error-free than that the least significant digits are received without error.

So the kind of error rate which is acceptable depends very much on the kind of message being sent.

In an instrumentation system the acceptable rate depends on the accuracy required. For speech an error rate of 1 in 10^5 (i.e. on average one digit in every 10^5 will be an error) is acceptable. As far as computers are concerned, it is usually the case that any uncorrected error at all is a disaster. It is not a question of a signal being more or less affected by the interference of noise, because any error will lead to a false instruction or a numerical error in the data and the accuracy for which the computer was being used will have been lost.

5.4 Discussion

The choice of modulation method appropriate to any particular instrumentation system depends upon a number of factors: accuracy, convenience, reliability, the channel bandwidth available, cheapness and so on. Each instrumentation system will have different requirements and problems, so it is impossible to say which is the 'best' modulation method. However, I have outlined a few of the differences between FM, AM and PCM, so that you are able to appreciate some of the factors which are involved in choosing a modulation method to be used over a noisy channel of an instrumentation system.

Self-assessment answers and comments

SAQ 1

A band-limited, time-limited, random waveform is one which cannot be specified with fewer than $2f_cT_o$ quantities, where f_c is the bandwidth and T_o is the duration of the waveform.

Deterministic waveforms of bandwidth f_c and duration T_o can, in general, be specified with fewer than $2f_cT_o$ quantities.

SAQ 2

(a) No, a random signal will take different values in different averaging intervals, so its average or mean value will be different during each interval.

(b) A stationary random signal is a random signal for which the parameters that describe it, for example, average value, r.m.s. value, do not depend on the time at which they are measured.

SAQ 3

The r.m.s. value of noise will be reduced by the averager by $\sqrt{400}$ relative to the r.m.s. value of the signal. The required ratio is therefore 20/1.

SAQ 4

If the output of one of the set of filters of the spectrum analyser is 8 V r.m.s. into 50 Ω and the bandwidth of the filter is 40 Hz, the power is $8^2/50=1.28$ W and the measured power density is

$$\frac{8^2}{40\times 50}=0.03 \text{ W Hz}^{-1}.$$

SAQ 5

The only parameter required to complete the characterization of the probability density function is the standard deviation of the waveform. Since the mean of the waveform is zero, the standard deviation is equal to the r.m.s. value of the waveform. Now, the power that the waveform dissipates in a known load resistor can be determined by finding the area under the power density curve. This

Power = (r.m.s. voltage)^2 ÷ value of the load resistor.

The value of the load resistor is known, so the r.m.s. voltage of the waveform can be found.

SAQ 6

The power that the sinusoid can dissipate in a resistor of resistance R before it passes through the filter is $(V_{in})^2/R$, where V_{in} is the r.m.s. value of the sinusoid at the input to the filter. For a sinusoid this r.m.s. value is $1/\sqrt{2}$ times its peak voltage $\hat{V}_{in}$. (Refer back to Unit 11 if you are unsure about this point.)

The power that the sinusoid can dissipate in the same resistor after passing through the filter is $(V_{out})^2/R$, where V_{out} is the r.m.s. value of the sinusoid at the output of the filter.

For the output power to be half the input power, $(V_{out})^2/R$ must be equal to $\frac{1}{2}(V_{in})^2/R$. This means that

$$\frac{(V_{out})^2/R}{(V_{in})^2/R}=\frac{1}{2}.$$

$$\frac{V_{out}}{V_{in}}=\frac{1}{\sqrt{2}}.$$

This ratio expressed in decibels is

$$\frac{V_{out}}{V_{in}}=20\log_{10}\frac{1}{\sqrt{2}}\text{ dB}$$

$$\approx -3\text{ dB}.$$

SAQ 7

The band-pass filter attenuates frequencies outside its bandwidth f_2–f_1, while those frequencies within its pass band are essentially unaffected.

The band-stop filter attenuates frequencies within a certain bandwidth f_2–f_1 and leaves frequencies outside the band essentially unaffected.

SAQ 8

Twelve sample values are given in Table 1 of Unit 11 to describe one period of the sine wave. They are

0.5, 0.87, 1.0, 0.87, 0.5, 0.0, −0.5, −0.87, −1.0, −0.87, −0.5, 0.0

The autocorrelation coefficients are determined as in the text.

First autocorrelation coefficient

signal	0.5	0.87	1.0	0.87	0.5	0.0	−0.5	−0.87	−1.0	−0.87	−0.5	0.0
zero shift	0.5	0.87	1.0	0.87	0.5	0.0	−0.5	−0.87	−1.0	−0.87	−0.5	0.0
products	0.25	0.75	1.0	0.75	0.25	0.0	0.25	0.75	1.0	0.75	0.25	0.0

Total of products = 6.0.
First autocorrelation coefficient = 0.5.

Second autocorrelation coefficient

signal	0.5	0.87	1.0	0.87	0.5	0.0	−0.5	−0.87	−1.0	−0.87	−0.5	0.0
1 shift	0.0	0.5	0.87	1.0	0.87	0.5	0.0	−0.5	−0.87	−1.0	−0.87	−0.5
products	0.0	0.44	0.87	0.87	0.44	0.0	0.0	0.44	0.87	0.87	0.44	0.0

Total of products = 5.24.
Second autocorrelation coefficient = 0.44.

Third autocorrelation coefficient

signal	0.5	0.87	1.0	0.87	0.5	0.0	−0.5	−0.87	−1.0	−0.87	−0.5	0.0
2 shifts	−0.5	0.0	0.5	0.87	1.0	0.87	0.5	0.0	−0.5	−0.87	−1.0	−0.87
products	−0.25	0.0	0.5	0.75	0.5	0.0	−0.25	0.0	0.5	0.75	0.5	0.0

Total of products = 3.00.
Third autocorrelation coefficient = 0.25.

Further coefficients can be calculated in a similar way.
Fourth autocorrelation coefficient = 0.0.
Fifth autocorrelation coefficient = −0.25.
Sixth autocorrelation coefficient = −0.44.
Seventh autocorrelation coefficient = −0.5.
Eighth autocorrelation coefficient = −0.44.
Ninth autocorrelation coefficient = −0.25.
Tenth autocorrelation coefficient = 0.0.
Eleventh autocorrelation coefficient = 0.25.
Twelfth autocorrelation coefficient = 0.44.

These twelve correlation coefficients are plotted in Figure 50 along with nine values for negative shifts (obtained by shifting the signal to the left).

You should be able to see from Figure 50 that the autocorrelation function of a sine wave is itself a sine wave with the same frequency in parametric time as the original signal has in real time.

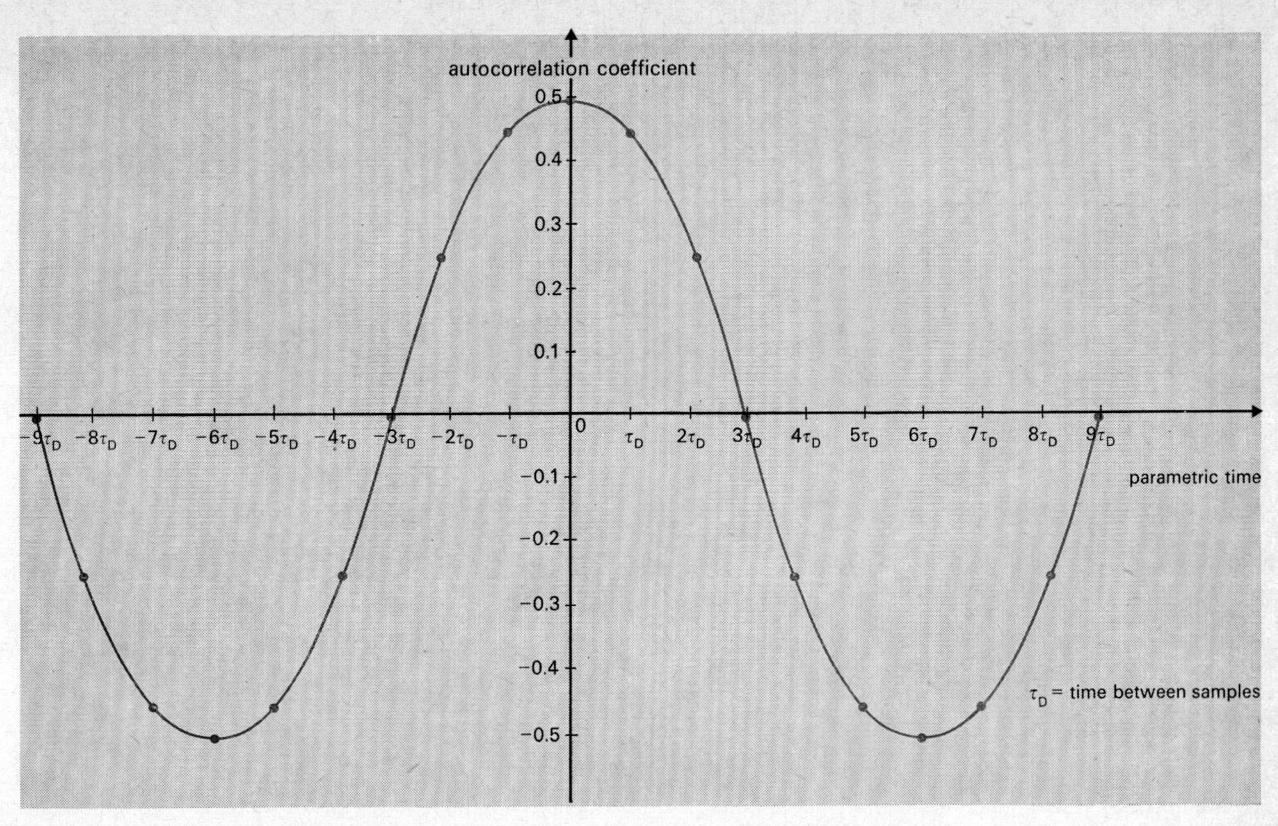

Figure 50 The autocorrelation function obtained from the sine-wave sample points given in Unit 11

SAQ 9

Parametric time −2

signal	−1	−1	+1	+1	+1	−1	+1	+1	−1	−1	+1	−1	+1	−1	−1
−2 shifts	+1	+1	+1	−1	+1	+1	−1	−1	+1	−1	+1	−1	−1	−1	−1
product	−1	−1	+1	−1	+1	−1	−1	−1	−1	+1	+1	+1	−1	+1	+1

Total = −1.

Average = $-\frac{1}{15}$.

Parametric time −1

signal	−1	−1	+1	+1	+1	−1	+1	+1	−1	−1	+1	−1	+1	−1	−1
−1 shifts	−1	+1	+1	+1	−1	+1	+1	−1	−1	+1	−1	+1	−1	−1	−1
product	+1	−1	+1	+1	−1	−1	+1	−1	+1	−1	−1	−1	−1	+1	+1

Total = −1.

Average = $-\frac{1}{15}$.

Parametric time 0

signal	−1	−1	+1	+1	+1	−1	+1	+1	−1	−1	+1	−1	+1	−1	−1
0 shifts	−1	−1	+1	+1	+1	−1	+1	+1	−1	−1	+1	−1	+1	−1	−1
product	+1	+1	+1	+1	+1	+1	+1	+1	+1	+1	+1	+1	+1	+1	+1

Total = 15.
Average = 1.

Parametric time +1

signal	−1	−1	+1	+1	+1	−1	+1	+1	−1	−1	+1	−1	+1	−1	−1
+1 shifts	−1	−1	−1	+1	+1	+1	−1	+1	+1	−1	−1	+1	−1	+1	−1
product	+1	+1	−1	+1	+1	−1	−1	+1	−1	+1	−1	−1	−1	−1	+1

Total = −1.

Average $= -\frac{1}{15}$.

Parametric time +2

Average $= -\frac{1}{15}$ (see Table 3 in text).

Parametric time +3

signal	−1	−1	+1	+1	+1	−1	+1	+1	−1	−1	+1	−1	+1	−1	−1
+3 shifts	+1	−1	−1	−1	−1	+1	+1	+1	−1	+1	+1	−1	−1	+1	−1
product	−1	+1	−1	−1	−1	−1	+1	+1	+1	−1	+1	+1	−1	−1	+1

Total = −1.

Average $= -\frac{1}{15}$.

Unit 13

Contents

Introduction

SAQ 1 (revision)

SAQ 1

A contiguous-band spectrum analyser X has 1 kHz bandwidth filters. Another contiguous-band spectrum analyser Y has 4 kHz filters. If both spectrum analysers are used to make measurements upon a signal with a uniform spectrum, what will be the ratio:

(a) of the power at the output of X's filters to the power at the output of Y's filters?

(b) of the power density readings obtained using X to the power density readings obtained using Y?

SAQ 2 (revision)

SAQ 2

What is the value of the autocorrelation coefficient, for a time shift equal to two sample intervals, for the waveform whose sample values are

+1 +1 +1 −1 −1 −1 +1 +1 +1 −1 −1 −1?

This unit continues with the question on which the last unit ended, namely: How can signals be processed and received in order to reduce the interference resulting from transmission through noisy channels?

Now, the two modulation methods, FM and PCM, considered in Unit 12 simply reduce the effects of noise. They do not allow the receiver to detect an error or correct it automatically.

Evidently, if noise has not been adequately removed from the received signal by filtering, averaging or modulation, for example, the wanted signal will still not be an accurate indication of the measurand. However, it is possible, as we shall see in a moment, to process a signal further so that the receiver of the signal is given an indication of when an 'error' in the signal has occurred and even to correct it during the process of reception.

These ideas of *error detection* and *error correction* are really only practicable at present in instrumentation systems if PCM or some similar digital signal is used, because the only signal errors which can satisfactorily be handled by simple electronic apparatus are digital errors. A digital error is the reception of a 1 when a 0 was transmitted or vice versa.

In some systems the receiver merely *detects* the presence of an error. In others, the receiver can be made to *correct* the received signal.

Both error-correcting and error-detecting capabilities can be achieved in a signal transmission and receiving system by appropriately processing the transmitted signal. This processing is given the general name *coding*, and it is the topic which we shall take up first in this unit. (This topic was introduced in the *Technology Foundation Course* in Unit 3 'Speech, Communication and Coding'.*)

coding

Later in the unit we shall consider the uses that can be made of the *cross-correlation function*, a function which is very similar to the autocorrelation function I talked about in Unit 12. Among other things, we

cross-correlation function

**The Open University (1972) T100* The Man-made World: Technology Foundation Course, *Open University Press.*

shall see that the cross-correlation function is the basis of an ingenious type of flow meter.

But first of all, let us consider coding. If you have taken the *Technology Foundation Course*, Section 2 will be partly revision, but I suggest you read it nevertheless.

Error-detecting codes

An error-detecting code is one which reveals to the receiver in some way that an error in transmission has occurred. It is rather like the situation when one person is talking to another and the listener says, 'What?' or, 'Could you repeat that?' The listener knew that a message had been sent but he was unable to decipher what that message was. Therefore he knew that an error had occurred somewhere along the channel between the speaker's intended signal and his decoding of the sound he received, so he asked for a retransmission. First, then, I want to consider what features there must be in the kind of signal which allows the receiver to decide that an error has taken place.

I shall only consider digital signals such as PCM – speech is much too complicated – so that the kind of errors I shall consider are simply errors in a binary signal, that is, the reception of '0' when '1' was transmitted, or the reception of '1' when '0' was transmitted.

The essence of all error-detecting or error-correcting techniques is the addition of *redundancy* to the signal. With a digital signal this means adding more digits to each binary signal in such a way that these extra digits are some function of the original signal and do not increase the set of symbols from which the message has been derived. This set of symbols is often called the *vocabulary* or *alphabet* of symbols.

redundancy

vocabulary alphabet

Let us take a very simple example, a four-digit natural binary code. In this code every four-digit pattern stands for something, and since there are sixteen different ways of arranging four binary digits, the alphabet or vocabulary of such a code contains sixteen 'words' or elements.

How big is the vocabulary of a six-digit binary code?

Just as there are $2^4=16$ words for the four-digit code, there are $2^6=64$ words for the six-digit one. (The 2 in the calculation arises because binary digits are being used; the index gives the number of digits.)

If we had to transmit the four-digit natural binary code for the number eight, namely 1000, along a noisy channel, it might be received as 1100 – an error having occurred in the second digit. This is the natural binary code for the number twelve. So the transmitted 'eight' will be received falsely as 'twelve'. Every four-digit pattern stands for something, so there is no redundancy in the code and every error in transmission causes an error in reception. Now, if the four-digit code was sent *twice* every time a number was required to be transmitted, then any single error could be detected. Thus, using this technique, if 'eight' were transmitted as [10001000] and received as [11001000], the difference between the two halves of the signal would reveal the presence of an error. But it would not reveal which half of the signal contained the error, since either half could contain the single error. So such a code, comprising the double or repeated transmission of each code word, can be called an error-detecting code. But it is a very inefficient one, as we shall see, for, although it involves transmitting four extra digits, it only enables the receiver to detect one error.

What combination of two errors would the receiver be unable to detect?

It would not detect the occurrence of an error in the same digit in each transmission of the word. For example, a transmission of [10001000] which was received as [11001100] would not give any error indication, it would just be received as if it were a correct transmission of the number twelve.

This simple code is, however, capable of detecting two, three or four errors provided they occur at different locations in the digit patterns. But this is of little value, for we need codes which detect errors wherever they occur. Better still, we need codes which enable the receiver to correct errors automatically and remove the need for asking for a repeat transmission of the error-affected signal.

Before we continue, let me just make clear the terminology which I have used in this introduction and which I shall continue to use throughout this unit.

The *message* is the information which is to be transmitted. It may be transmitted directly, it may be used to modulate a carrier or it may be coded. A convenient part of this message is called a *message word*. The number thirteen, for example, is a 'word' in this sense. It might be the quantized value of an instrument reading. The *vocabulary* is the complete set of words used in the PCM coding of the messages.

message

message word

vocabulary

Each word of the message may be coded using natural binary code or a Gray code. This is called a *code word*. Thus 1101 is the natural binary code word for the message 'thirteen'.

code word

If the message is further coded for error detection or error correction by the addition of extra digits, so as to produce redundancy, the resulting sequence of digits is also called a *code word*. It may be necessary to distinguish between code words which contain no *redundancy* (e.g. Gray code or natural binary) and redundant codes with error-detecting properties. Where this is necessary I shall use the phrase *message code word* to describe the non-redundant pattern and the phrase *redundant code word* when there are more digits in the signal than are strictly necessary to encode the message.

message code word

redundant code word

Thus, using the simple error-detecting code described previously, the *message word* was 'eight', the *message code word* was 1000 and the *redundant code word* was 10001000.

Now let us look at some rather more efficient error-detecting codes.

2.1 Parity check codes

The simplest error-detecting code in common use involves the addition of one further digit, as mentioned briefly in Unit 7, where an extra digit was added to each BCD code when it was punched on paper tape. This means using *five* digits to transmit the numbers from zero to nine. Since, with five digits, it is possible to encode thirty-two different numbers, over half the possible code patterns are not used when a message vocabulary of only ten (zero, one, two, . . ., nine) is all that is required. If one of the unused patterns arrives at the receiver, it must be the case that an error has occurred.

The extra digit in this system is called a 'parity check digit'. In an *even-parity* system the extra digit is added in such a way as to ensure that there is an even number of 1s in any correct code. Thus, any received pattern with an odd number of 1s in it must be in error.

Alternatively, *odd parity* can be used, in which those code words with an odd number of 1s are correct and those with an even number are in error.

Table 1

Message code	Code word with parity digit added
0101	01011
0111	01110
0100	01000
1010	10101

In Table 1 is it even parity or odd parity that is being used in the construction of the code words?

Odd parity. All the five-digit patterns have an odd number of 1s in them.

This technique of adding parity check digits is widely used in computers and numerically controlled machine tools and makes possible an automatic indication of the occurrence of an error.

SAQ 3

SAQ 3

For the automatic detection of errors in parity check codes a circuit is needed to detect odd or even parity. Devise a four-input logic circuit using EXCLUSIVE-OR gates (or just NAND gates) whose output is 1 when (and only when) there is an *odd* number of 1s applied to the inputs.

Note, by way of revision. An EXCLUSIVE-OR gate is a two-input logic circuit which gives an output of 1 when the inputs are different (1 and 0, or 0 and 1) but gives an output of 0 when the inputs are the same (0 and 0 or 1 and 1). It can be constructed from NAND gates as shown in Figure 1.

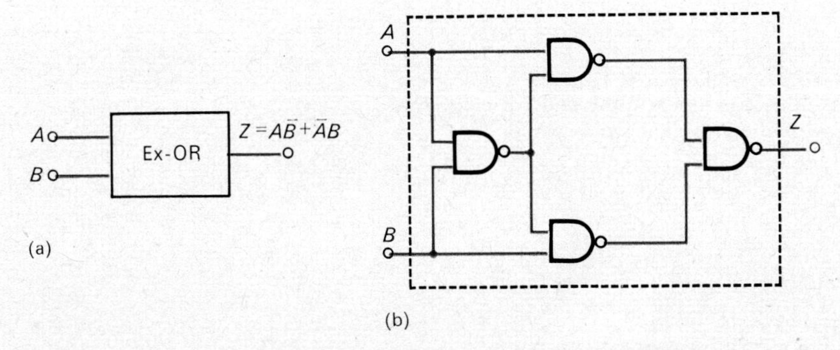

Figure 1 The EXCLUSIVE-OR gate: (a) as a circuit element with the stated logic function; (b) how it can be made from NAND gates

These simple parity check codes, either even or odd, are unable to detect reliably more than one error. Two errors will always produce an acceptable signal code word. For example, consider the first code word in Table 1, 01011. Suppose the first error occurs in the first digit to give 11011, the second error can then occur in any of the four remaining digits to give 10011, 11111, 11001 or 11010. Each of these has odd parity and would therefore be recognized as a valid code word.

In order to detect more than one error it is necessary to add more digits to the message code.

2.2 The ARQ code

The most commonly used error-detecting code in telegraphy is the ARQ code or the 'four out of seven' code. Each code word consists of four 1s and three 0s (e.g. 1010101 or 1111000). Strictly speaking, this code has nothing to do with instrumentation, but it is offered as an example of a simple but powerful coding technique.

In this case each message code word is not *added to* in orde1 to produce a redundant code word. Instead the message is coded into a new redundant form. There are thirty-five different arrangements of four 1s and three 0s and these are used to stand for the twenty-six letters of the alphabet, plus various instructions, etc. But there are 2^7 (=128) ways of arranging seven binary digits, so the *message code words* contain quite a lot of redundancy.

SAQ 4

SAQ 4

Show that there are thirty-five ways of arranging four 1s and three 0s in a sequence of seven digits.

(You can, of course, just write down all possible arrangements, but with a little thought you might be able to work out that there must be thirty-five arrangements even though you do not write down any. One way of doing this is to relate this question to the problem dealt with in Unit 11 of making a nominally 70 Ω resistor from 9 Ω and 11 Ω resistors.)

On reception of each ARQ code pattern the task of the receiver is to check that the pattern contains just four 1s and three 0s. If it does not, then an error has occurred and the receiver Automatically ReQuests retransmission (hence the name ARQ). If two errors occur, there is, of course, a chance that there will be four 1s and three 0s in the received pattern despite these two errors; one of the 1s can be received as an 0 and one of the 0s can be received as a 1. Thus this code cannot be relied upon to detect more than one error.

SAQ 5

SAQ 5

About what proportion of double errors will the code detect? (*Hint.* List all the combinations of two errors that can occur.)

A None;
B 15 per cent;
C 50 per cent;
D 85 per cent.

A simple error-detecting code, such as the ARQ, works well in a practical system so long as not too many errors occur. If the probability of an error is rather high, so that the retransmission of a signal is also likely to contain an error, it is easy to see that progress can be seriously held up by frequent requests for retransmission.

Error-correcting codes

A code which allows automatic error *detection* to take place is of only limited value. But a code which automatically *corrects* any errors which occur in a code word, up to a certain maximum number, avoids the necessity for retransmission when errors occur in smaller numbers. It also leads to an improvement in the accuracy of signal transmission.

> Suppose that in a PCM channel there is said to be a 1 in 10^5 chance of an error occurring as a result of interference by Gaussian noise in the transmission channel. What does this mean?

It means that, *on average*, one digit in 10^5 digits will be received in error. So, if each code word transmitted consists of five digits then, on average, one word in $10^5 \div 5$, that is, 1 in 20 000, received words will be in error to the tune of one digit. Alternatively, it means that if each code word contains ten digits then, on average, one word in 10 000 will contain at least one error. In either case some code words will contain two errors.

Let us now suppose that by the addition of five extra digits to a five-digit code word we can make it capable of correcting any single error. Also let us assume that the chance of *two* errors occurring in a ten-digit word is 1 in 10^6. We now have two possibilities:

1 If we send the signal unmodified as a sequence of five-digit code words, one word in 20 000 words will be in error.

2 If we send the same five-digit messages in the form of a ten-digit error-correcting code, we increase the chance of a single error in each code word to 1 in 10 000, but we now have only to worry about double errors because all single errors are corrected. Double errors occur only once in a million transmissions, so this is a fiftyfold improvement.

Although increasing the length of a code word increases the chance of an error occurring in the word, the error-correcting capability can give an overall reduction in the probability of incorrectly *decoding* (reconstituting the message signal from the received signal) the code.

decoding

If, however, double errors were almost as common as single errors, correction of single errors would not be helpful – indeed it could increase the number of errors.

> Why?

Because it increases the length of the code word, and so it increases the chance of two errors occurring within the word.

The whole subject of error-correcting codes is one with a deep mathematical theory underlying it.* The basic principles can, however, be understood from considering a simple example. So I shall be describing one kind of error-correcting code – the *chain code*. The reasons for my choice are as follows:

chain code

**If you are interested in these mathematical aspects and wish to go into them further, you might refer to a textbook such as Rosie, A.M. (1973)* Information and Communication Theory, *Van Nostrand.*

1 You are already familiar with some of the properties of chain codes, since chain code is another name for the pseudo-random sequence that we have already discussed in Unit 12 and that you may have studied in the *Technology Foundation Course*.

2 The way in which chain codes can provide error correction is quite easy to understand.

3 Chain codes are being successfully used in the transmission of instrument readings from space probes to Earth, and they are a key part of some instrumentation systems today.

First I shall describe how they are generated, then how they can be decoded and finally how their properties allow the automatic correction of errors.

3.1 Chain codes

As explained earlier, the essence of any error-correcting code is the addition of extra digits to the original message code so that their presence can be used to reveal where errors in transmission have occurred. These extra added digits are *redundant* digits because they do not increase the number of possible message words that can be sent. If all is going well and there is no noise in the channel, they are not needed. They are only needed when errors occur. The method of construction of a chain code from a given message code is similar for message codes of different lengths, but let us first go through the process for a four-digit message code word.

You can construct a chain code from a four-digit message as follows:

1 Start with your message code word, say 1101 (any four-digit pattern except 0000; I shall return to this rather special case a little later on).

2 Add an extra digit to the end of the four digits according to the rules:

(a) If the first two digits are the same, add a 0 to the end of the four digits.
(b) If the first two digits are different, add a 1 to the end of the four digits.

These rules for deciding whether to add a 0 or 1 to the end of the digit pattern are called *modulo-2 addition*.

modulo-2 addition

(Thus, starting with 1101, we add a 0, because the first two digits are the same, forming the pattern 11010.)

3 Discard the first digit, take the last four digits of the new five-digit pattern and repeat steps 2 and 3 until the original message signal is about to be regenerated.

The redundant code word is the sequence of initial digits that are discarded in performing the first part of step 3.

This procedure should become clear with the help of an example.

Example

Start with the message code word 1101. Add an extra digit according to rule 2 above and drop the first digit, giving 1010. Repeat this procedure until the original message code word is formed again.

1101
1010
0101
1011
0111
1111
1110
1100
1000
0001
0010
0100
1001
0011
0110
——
1101 original message signal

The chain code is then the sequence of digits formed by taking the first digit of each row; namely, in this case,

110101111000100. (A)

SAQ 6

SAQ 6

Generate the chain code corresponding to a four-digit message 1000. What do you notice about your chain code compared with the code for 1101?

You *could* say that the first four digits are the message and the next eleven digits are error-correcting digits, but we shall see that this is not a helpful description because in order to decode the signal the whole code word is dealt with. It is not divided into the message plus error-correcting digits.

The precise rule for the construction of the extra digits depends upon the length of the original message word.

For your information only, the rules for constructing chain codes for various lengths are:

Foı a three-digit message word you perform modulo-2 addition on the first and third digits to produce each new digit. The resulting chain code is seven digits in length.

For a four-digit message word, as explained, you perform modulo-2 addition on the first two digits to produce each new digit. The resulting chain code is fifteen digits in length.

For a five-digit message you perform modulo-2 addition on the first and third digits to produce each new digit. The resulting chain code is thirty-one digits in length.

For a six-digit message you perform modulo-2 addition on the first two digits to produce each new digit. The resulting chain code is sixty-three digits in length.

For a seven-digit message you perform modulo-2 addition on the first and fifth digits to produce each new digit. The resulting chain code is 127 digits in length.

These chain codes may be produced using an electronic circuit consisting of a *shift register* and an EXCLUSIVE-OR gate. You met the

EXCLUSIVE-OR gate earlier in the unit (SAQ 3) so, for the moment, let me briefly explain the action of the shift register.

A shift register consists of several *JK* bistables connected together as shown in Figure 2(a). On application of a clock pulse to the bistables the pattern shifts one place to the right, as shown in Figure 2(b). (You may like to prove to yourself that this will happen by applying the rules concerning the operation of *JK* bistables given to you in Unit 6.) The digit 1 that was on the extreme right of the circuit has now been shifted out of the register and the digit 0 that was the input prior to the clock pulse is now stored on the first bistable Q terminal.

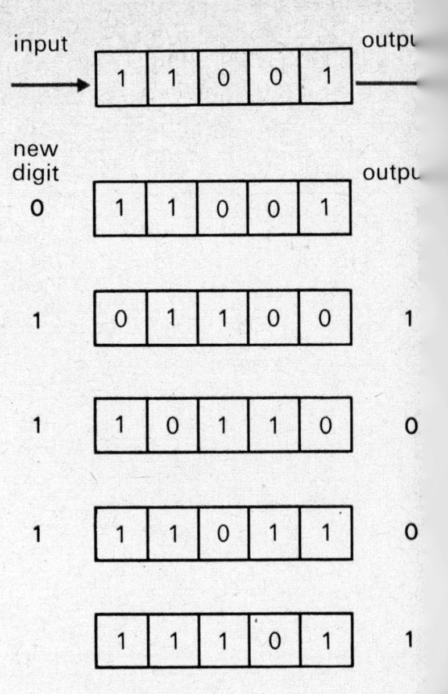

Figure 3 *A schematic diagram of a five stage shift register containing the digi 11001 and the action of the shift registe for five successive clock pulses*

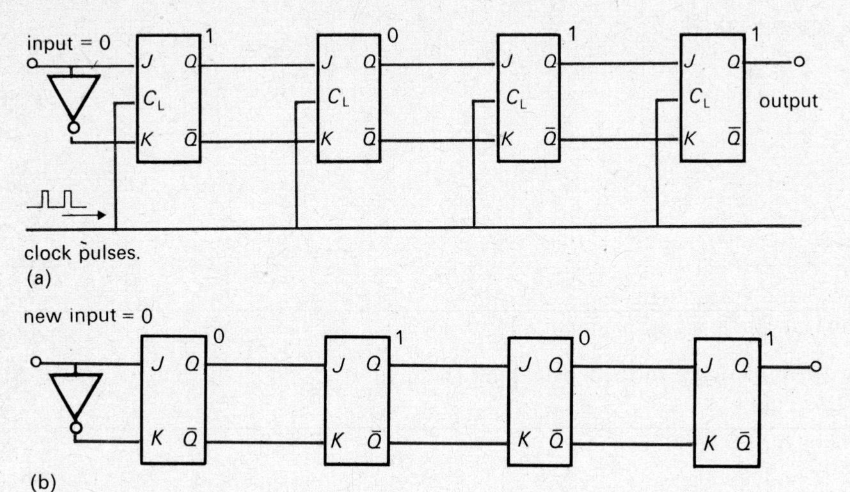

Figure 2(a) Four JK bistables connected together to form a four-stage shift register. (b) The state of the bistables after the application of a clock pulse

You can construct a simple shift register using your logic tutor. The details are given in your Home Experiment Book. (You should attempt this exercise *only if you are interested and can spare the time.*)

The shift register is illustrated schematically in Figure 3, which also uses this representation to illustrate the action of a shift register for five successive clock pulses. (The 'new digits' were chosen quite arbitrarily just to illustrate the action of the circuit.)

To produce a chain code the digits in the message code word can be entered into the shift register one by one by presenting them at the input bistable prior to each clock pulse, as shown in Figure 4. Next an EXCLUSIVE-OR gate is connected to the shift register so as to perform modulo-2 addition on the contents of the appropriate stages of the shift register.* The output of the EXCLUSIVE-OR gate is connected to the input of the shift register. The connection of the EXCLUSIVE-OR gate to a four-stage shift register is shown in Figure 5.

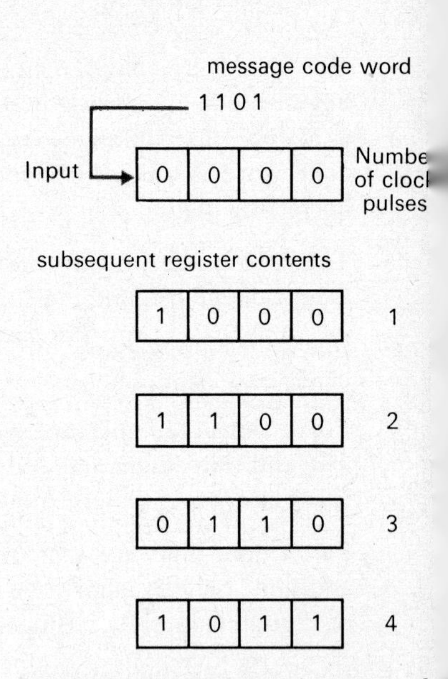

Figure 4 The procedure for putting th message word into the shift register

Further clock pulses will now cause the complete chain code word to be generated as a sequence of states of the last stage of the register.

In your Home Experiment Book you will find an exercise on the generation of a chain code using your logic tutor. *If time permits*, you may like to attempt this exercise.

SAQ 7

Generate the chain code for the three-digit message word 111.

SAQ 7

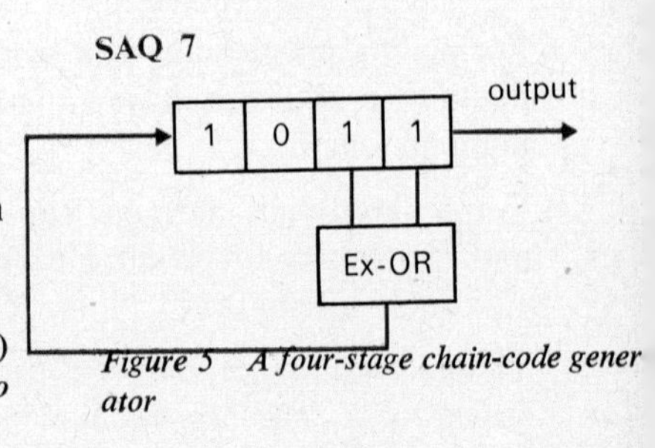

Figure 5 A four-stage chain-code gener ator

So much for the generation of a chain code corresponding to a given message word, now let us look at the properties of a chain code.

**Compare the rule for modulo-2 addition given at the beginning of this section (3.1) with the operation of the EXCLUSIVE-OR gate described in the revision note to SAQ 3.*

3.2 The properties of chain codes

3.2.1 *The characteristic code sequence*

You should have noticed that the chain code you constructed in SAQ 6 was the same sequence of digits as given in sequence (A) in the example, but *shifted along a few digits*. Indeed, if we draw the sequence shown in (A) as a circle with the end joined to the beginning, representing the fact that (A) repeats itself, then this endless chain tells you the code pattern for all possible four-digit message codes (except 0000), as shown in Figure 6. First find the four-digit message code you want to encode, then the chain code word is the sequence of digits clockwise around the circumference of the circle, starting from the message code word. The chain code for 1101 is 110101111000100 (sequence A), while for 1111 it is 111100010011010.

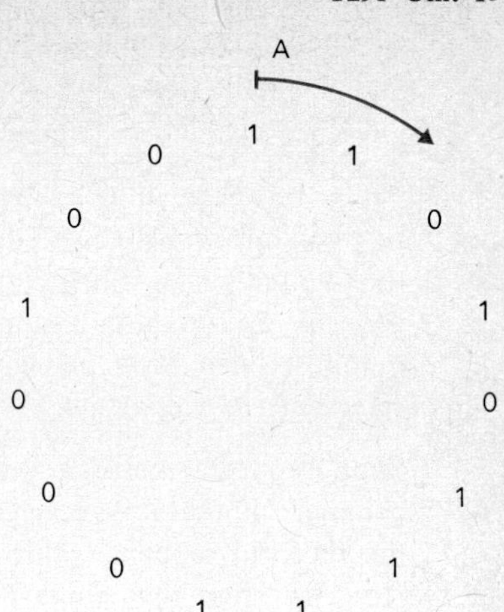

Figure 6 The fifteen-digit chain code arranged in a circle

So the chain code words are all the same cyclic sequence, or chain, of digits, and the feature which distinguishes one code word from another is the starting point in the chain (hence the name 'chain code').

3.2.2 *The length of the code*

You saw, when constructing the code for a four-digit message, that the number of digits in the whole sequence was equal to the number of lines in the construction table. But each line corresponded to one possible arrangement of four digits. Every possible arrangement of four digits except 0000 appears at the beginning of a line, and since there are a total of 2^4 possible arrangements of four digits, the number of digits in the whole sequence must be $2^4-1=15$. In general, the length of a chain code is 2^n-1, where n is the number of digits in the message word.

3.2.3 *Pseudo-random characteristics*

The sequences of 1s and 0s comprising the chain codes are also called *pseudo-random binary sequences.* The sequence (A) is the same sequence referred to in Unit 12 as a pseudo-random binary sequence. The reason for this name being applied to the chain codes is that a chain code sequence such as (A) has many properties similar to those of random binary signals. Notice that for a code sequence:

1 The occurrence of 1s and 0s is aperiodic. This is also true for a portion of a random binary signal.

2 There is always a nearly equal number of 1s and 0s. Actually there is one more 1 than there are 0s. (A random binary signal, with zero mean would have, on average, an equal number of 1s and 0s.)

3 The autocorrelation function of the sequence is a 'spike'. (We discussed this in Unit 12.)

3.2.4 *Error-correcting capability*

We have seen that each sequence of a chain code has the same ordering of 1s and 0s. The only difference is that each of the sequences begins at a different point.

Now there is another remarkable property of chain codes which establishes the error-correcting capability. It is this. If you take any two chain codes and add them together using modulo-2 addition you obtain another version of the same chain code. The following example shows what I mean.

The first two rows in Table 2 are two fifteen-digit chain codes.

Table 2

Code for 1101	1	1	0	1	0	1	1	1	1	0	0	0	1	0	0
Code for 1111	1	1	1	1	0	0	0	1	0	0	1	1	0	1	0
Sum	0	0	1	0	0	1	1	0	1	0	1	1	1	1	0

=code for 0010

The third row is their sum. Remember that in modulo-2 addition $1+1=0$ and $0+0=0$, while $1+0=1$ and $0+1=1$. The sum is the same sequence again, and contains eight 1s and seven 0s.

Note, however, that each 1 in the sum corresponds to a difference between one sequence and the other. So there are *eight differences* between the two sequences. Furthermore, since you get the same result (i.e. a chain code as the modulo-2 sum) whatever two chain codes you start with, it follows that *there are eight differences between any fifteen-digit chain code and every other one.*

Suppose you receive a sequence after transmission through a noisy channel and that it is received with, say, two errors in it. You can compare the received sequence with every one of the possible correct sequences. You will find that there are two differences with one of the correct sequences, but at least six with every other one – so you could decide which sequence must have been sent.

If there were three errors, could you tell which was the correct sequence?

There would be at least five errors with all but the correct sequence, so you could still decide. If there were four errors, however, you might be uncertain because there could be two sequences differing by four digits from the received sequence. Thus, by a process of comparison between the received signal and all possible correct ones, you can correct up to three errors.

SAQ 8

SAQ 8

How many errors can you correct in a chain code of thirty-one digits?

The most important property of the chain code for our purpose is that it is an error-correcting code. Let us see how it is possible in practice to use chain codes to correct errors.

Decoding a chain code

Chain codes are used to transmit data over noisy channels whose bandwidth is usually just sufficient to accommodate this signal. (This bandwidth depends on the rate of transmission of the individual digits in the chain code word.)

Figure 7(a) shows a portion of a transmitted chain code word and Figure 7(b) shows what the received signal corresponding to this portion of the code word would typically look like at the receiver after transmission over such a channel.

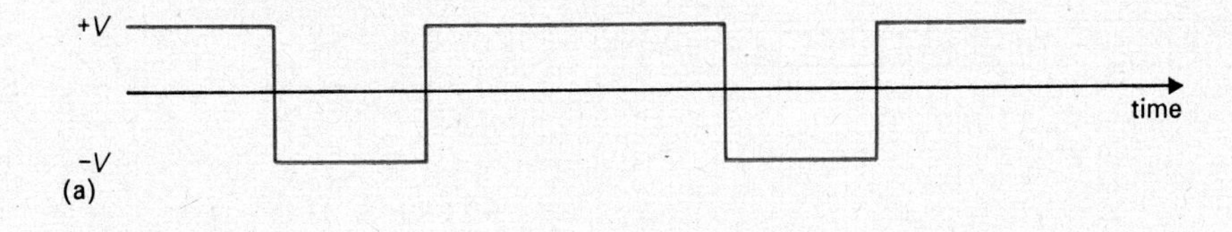

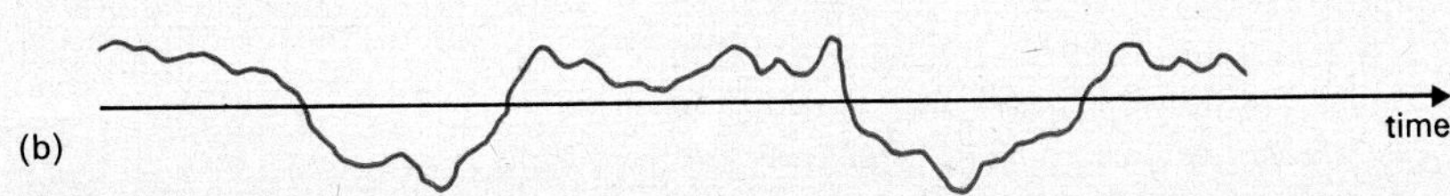

Figure 7(a) A segment of a chain code. (b) The received segment after transmission over a noisy band-limited channel

One method of decoding a chain code word involves first deciding digit by digit if the received waveform represents a 1 or a 0. (One way of doing this was described in the radio programme 'Signal statistics'.) In this way the received signal corresponding to a chain code word is converted into a sequence of 1s and 0s. If all the decisions are made correctly, this sequence is said to be *error free* and it will be identical to the transmitted sequence. But noise added to the signal can cause the receiver to make incorrect decisions and the received sequence will then contain errors.

However, as we shall see, provided there are not too many errors the received sequence will be decoded correctly.

The next step in the decoding of a received chain code is to obtain the message code word from the received sequence of digits. Usually we cannot be sure that no errors have been made in receiving the sequence of 1s and 0s, so we cannot say that it is given by the initial digits in the chain code sequence. We must use another method.

One method of doing this is to use a process very analogous to calculating the autocorrelation coefficient of a signal, but instead of calculating the autocorrelation coefficient between a signal and itself – shifted in time – we calculate the *cross-correlation coefficients* between the received signal and each one of all the possible correct chain codes in turn. These cross-correlation coefficients are calculated using a very similar procedure to that used in calculation of autocorrelation coefficients.

cross-correlation coefficient

Remember that the highest-value autocorrelation coefficient is 1 and it is always obtained for zero time shift. Of course, for zero time shift the waveform and the time-shifted waveform are identical.

Similarly, when decoding a chain code, the highest-value cross-correlation coefficient will be 1 and it will be obtained between a chain code word received error free and the stored version of that chain code word.

If the maximum value of cross-correlation coefficient is less than 1, this indicates that some errors are present in the received code word. But this maximum value of cross-correlation coefficient will indicate which of the stored code woıds is most like the received one.

One way of visualizing the comparison between the received code word and each of the stored code words is shown in Figure 8. The complete set of stored chain code words can be represented by one sequence. The process of calculating the cross-correlation coefficient between the received code word and each of the stored code words can be thought of as comparing the received code word and the stored sequence, starting from a different point in the sequence for each coefficient.

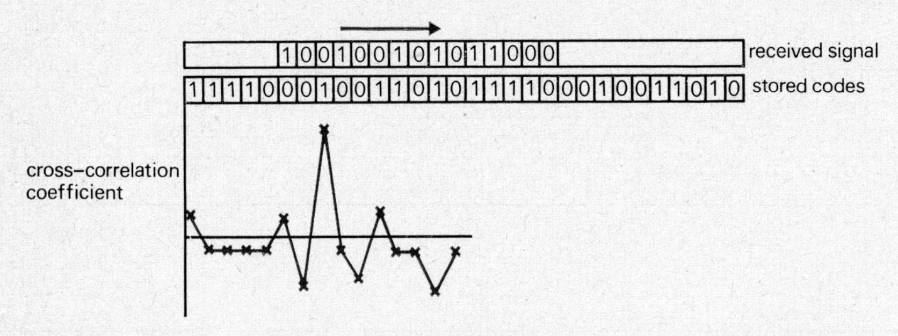

Figure 8 Detection of a chain code which includes errors. A received fifteen-digit sequence is compared by computing the cross-correlation coefficient with each one of fifteen possible error-free codes. The largest peak on the resulting graph of correlation coefficients indicates the location of the start of the error-free sequence. In the case shown the error-free sequence is 100110101111000, so we can deduce that there were two errors in the received signal and the message signal was 1001

The values of the cross-correlation coefficients obtained can be plotted against the starting point in the sequence, as shown in Figure 8. This graph is the *cross-correlation function* between the received code word and the stored sequence. I shall be talking more about this function later in the unit but for the moment all we need to be able to do is to calculate cross-correlation coefficients.

To make it clear how the calculation of cross-correlation coefficients enables us to decode a chain code, let us consider some examples. First we shall consider the decoding of a chain code which is received error free and then we shall consider the decoding of code words containing errors.

Suppose the transmitted signal is the chain code sequence

1 1 0 1 0 1 1 1 1 0 0 0 1 0 0 (A)

and it is received without error.

Let us now see how these calculations are performed. As in Unit 12, when I discussed autocorrelation, I shall represent 1 by $+1$ and 0 by -1. If the received code word (A) is compared with the same code word stored in the receiver, the cross-correlation coefficient is calculated as shown in Table 3.

Each digit of the received signal is multiplied by the corresponding digit of the stored signal and the cross-correlation coefficient is the average of all these products, $+1$ in this case.

Here the correlation of the received signal with the stored signal is the same as the autocorrelation coefficient of the signal for delay time $\tau=0$.

If the transmitted signal (A) is now compared with a different stored code, we shall obtain, for example, the products shown in Table 4.

Table 3

Received code	+1	+1	−1	+1	−1	+1	+1	+1	+1	−1	−1	−1	+1	−1	−1
Stored code	+1	+1	−1	+1	−1	+1	+1	+1	+1	−1	−1	−1	+1	−1	−1
Products	+1	+1	+1	+1	+1	+1	+1	+1	+1	+1	+1	+1	+1	+1	+1

The average of the products is +1.

Table 4

Received code	+1	+1	−1	+1	−1	+1	+1	+1	+1	−1	−1	−1	+1	−1	−1
Stored code	+1	+1	+1	+1	−1	−1	−1	+1	−1	−1	+1	+1	−1	+1	−1
Products	+1	+1	−1	+1	+1	−1	−1	+1	−1	+1	−1	−1	−1	−1	+1

Cross-correlation coefficient is $-\frac{1}{15}$.

The products comprise seven +1s and eight −1s, so the cross-correlation coefficient is equal to $-\frac{1}{15}$ in this case.

This is always the case if the cross-correlation coefficient is calculated for any two different error-free, fifteen-digit chain code words. You may like to check that this is always true by taking two such chain code sequences, finding the cross-correlation coefficient and examining the result. Repeat this procedure until you are convinced.

Thus, the correlation coefficient with the correct stored pattern is +1 *and with all others it is* $-\frac{1}{15}$, *if there are no errors in the received signal.*

This statement is also true for the code word produced from the message word 0000. This code word consists of fifteen zeros.

The cross-correlation coefficient between this sequence of zeros and any other chain code word is $-\frac{1}{15}$. Table 5 shows an example.

Table 5

Received code	−1	−1	−1	−1	−1	−1	−1	−1	−1	−1	−1	−1	−1	−1	−1
Stored code	+1	+1	−1	+1	−1	+1	+1	+1	+1	−1	−1	−1	+1	−1	−1
Products	−1	−1	+1	−1	+1	−1	−1	−1	−1	+1	+1	+1	−1	+1	+1

Cross-correlation coefficient is $-\frac{1}{15}$.

The cross-correlation coefficient between the all-zero code and a stored version of itself is, of course, +1.

Therefore, this code word behaves in the same way as the chain code words when it is used to calculate cross-correlation coefficients as described above. However, the use of this sequence as a code word has several practical disadvantages.

Before considering received signals containing errors I shall summarize the situation so far, in which the received signal contains no errors.

1 Starting with any four-digit message we can generate a chain code of fifteen digits.

2 Each chain code is the same sequence of digits but starting at a different point in the chain.

3 Each code contains eight 1s and seven 0s in an aperiodic sequence. The sequences are alternatively called pseudo-random binary sequences.

4 Detection of the original message can be achieved by comparing the received sequence with each possible sequence. This means calculating the cross-correlation coefficients between the received code sequence and each of the other chain code words which are stored at the decoder in some way.

5 For this error-free situation a cross-correlation coefficient equal to 1 will be obtained between the received code word and the identical stored code word.

6 Each error-correcting code word (for a fifteen-digit chain code) has a cross-correlation coefficient of $-\frac{1}{15}$ when compared with any of the other error-correcting code words.

4.1 The correction of errors using a chain code

Suppose there is an error in the received signal, as shown in the sequence in Table 6. (The error is indicated by bold type.)

Table 6

Received code	+1	+1	−1	+1	−1	**−1**	+1	+1	+1	−1	−1	−1	+1	−1	−1
Stored code	+1	+1	−1	+1	−1	+1	+1	+1	+1	−1	−1	−1	+1	−1	−1
Products	+1	+1	+1	+1	+1	−1	+1	+1	+1	+1	+1	+1	+1	+1	+1

Cross-correlation coefficient is $\frac{13}{15}$.

When compared with the correct stored code the products are now no longer all +1s, one of them is −1, so the cross-correlation coefficient falls to $+\frac{13}{15}$, instead of +1.

The same faulty received signal compared with any other of the stored signals may convert one of the −1s in the product to a +1, as indicated in Table 7.

Table 7

Received code	+1	+1	−1	+1	−1	**−1**	+1	+1	+1	−1	−1	−1	+1	−1	−1
Stored code	+1	+1	+1	+1	−1	−1	−1	+1	−1	−1	+1	+1	−1	+1	−1
Products	+1	+1	−1	+1	+1	+1	−1	+1	−1	+1	−1	−1	−1	−1	+1

Cross-correlation coefficient is $+\frac{1}{15}$.

Alternatively, one of the +1s in the product may be converted to a −1, as shown in Table 8.

The largest possible cross-correlation coefficient obtained by comparing the received code containing one error with an incorrect code word occurs when a product of −1 is changed to a +1 giving a coefficient of $+\frac{1}{15}$ instead of $-\frac{1}{15}$. This is still much less than the coefficient obtained by comparison with the correct code.

So the receiver which decides that the transmitted code word is the

Table 8

Received code	+1	+1	−1	+1	−1	−1	+1	+1	+1	−1	−1	−1	+1	−1	−1
Stored code	−1	+1	−1	+1	+1	+1	+1	−1	−1	−1	+1	−1	−1	+1	+1
Products	−1	+1	+1	+1	−1	−1	+1	−1	−1	+1	−1	+1	−1	−1	−1

Cross-correlation coefficient is $-\frac{3}{15}$.

one which gives the largest cross-correlation coefficient with the received code word can tolerate one error and still make the right decision. But what would happen if there were more errors in the received signal?

For instance, what is the cross-correlation coefficient between a received code containing three errors and:

(a) the correct code sequence;
(b) the incorrect code sequences?

(a) In this case two more +1 products will become −1s. The average of the products will become $+\frac{9}{15}$.

(b) The largest value of cross-correlation coefficient is obtained if three −1 products become +1s. This gives a coefficient of $\frac{5}{15}+$.

The correct code still gives a higher correlation than an incorrect code, so the decoder can tolerate three errors. Only when there are four errors in the in-coming signal is there a possibility of ambiguity as to which stored sequence the received signal most closely resembles. With five or more errors the decoder makes an incorrect decision.

For four errors there will be two stored code words giving cross-correlation coefficients of $\frac{7}{15}$: one correct and one incorrect. So the decoder cannot decide which code was transmitted. But it can *detect* the presence of four errors, since two correlations give $\frac{7}{15}$. Otherwise, for three errors or less, the highest cross-correlation coefficient reveals the stored code word that is nearest to the received one and is evidently then the correct code word.

SAQ 9

SAQ 9

Check that the highest cross-correlation coefficient obtainable with any received code with four errors in it is $\frac{7}{15}$.

Thus, so long as there are no more than three errors in the received code sequence, the cross-correlation coefficient between the received code and an error-free version of it is always higher than the correlation coefficient between the received code sequence and an error-free version of any other code word.

So, summarizing, to decode the signal we calculate the cross-correlation coefficients between the received signal and each possible error-free code word.

The coefficient which is $\frac{9}{15}$ or more indicates to which error-free code the received code sequence is nearest. If fewer than four errors have occurred in transmission, this code will be the transmitted code word.

The number of errors that any chain code can correct is obviously related to its length. The longer the chain code, the more redundancy it has and the more errors it can correct.

4.2 Error-correcting capabilities of a chain code

For a received error-free chain code of length L the cross-correlation coefficient with the stored version of the code word is $+1$ and with all other stored code words it is $-1/L$. Each single error will cause a reduction of $2/L$ in the cross-correlation coefficient with the correct word (with $+1$ being changed to -1 in the products sequence) and an increase of the same amount with an incorrect code word. The decoder can correct errors until the correlation with the correct code word is reduced to halfway between 1 and $-1/L$, that is, $\frac{1}{2}(1-1/L)$. In the previous example this was $\frac{1}{2}(1-\frac{1}{15})=\frac{7}{15}$. Then the decoder can only indicate the *presence* of errors.

If the correlation with the correct code word falls below this value, the decoder starts to make the wrong decisions.

The number of errors N_D that it takes to reduce the correlation with the correct code word to $\frac{1}{2}(1-1/L)$ is given by

$$N_D=\frac{1-\frac{1}{2}(1-1/L)}{2/L}$$

$$=\frac{L+1}{4}, \qquad (1)$$

since each error reduces the 'correct' cross-correlation coefficient by $2/L$. The length of the chain code was given in terms of n, the number of digits in the message word, in section 3.2.2 as $L=2^n-1$.

So the maximum number of errors N_D that the chain code can detect as given in equation (1) is

$$N_D=\frac{2^n-1+1}{4}=2^{n-2} \qquad (2)$$

and the maximum number of errors that the code can correct is one less than this number,

$$N_C=N_D-1=2^{n-2}-1. \qquad (3)$$

In the previous example $n=4$, $N_D=2^{n-2}=4$ and $N_C=2^{n-2}-1=3$.

Other methods have been devised for decoding these chain codes, but it is not possible to achieve greater error correction than this.

4.3 Discussion

Let me now try to fit these coding ideas into an instrumentation system.

We have the problem of conveying the results of some measurements through a noisy channel using error correction. What are the various steps in the process?

First, we convert the results of the measurements into binary code. If the transducer produces a continuously varying output, the binary code will represent the values of quantized samples of the signal. Thus, the signal is represented as a sequence of binary numbers.

Let us suppose that these are five-digit numbers corresponding to thirty-two quantized values and that samples are being taken at a rate of 100 samples per second. The overall digit rate is therefore 500 binary digits per second. The term 'binary digit' is often shortened to *bit*, so this digit rate is 500 bits per second. The next step is to produce a chain code for each five-digit sample value. Each code sequence is thirty-one digits in length, so now, in order to transmit the sample values at 100 per second, the bit rate must increase to 3100 bits per second.

bit

This sequence of chain codes of thirty-one digits each is transmitted and the receiver has the task of deriving from it the original sequence of sample values.

This, as we have seen, can be done by finding cross-correlation coefficients. But an essential part of this process is ensuring that the transmitter and receiver remain 'in step', so to speak. For error correction to be realized, the receiver must have an indication of when each code sequence begins. Thus, in addition, a synchronizing sequence has to be sent. This may take many forms. The transmission of an easily recognizable digit sequence such as 101010 . . . at regular intervals is one simple way.

Having determined which sequence of thirty-one digits is being sent in each sample period, the receiver retains just the first five digits, since these are the original message code for the sample value. Each five-digit code is passed through a digital-to-analogue converter to regain the original signal. (You will meet digital-to-analogue circuits at Summer School and again in Unit 14.)

See Figure 9 for a diagrammatic representation of these operations.

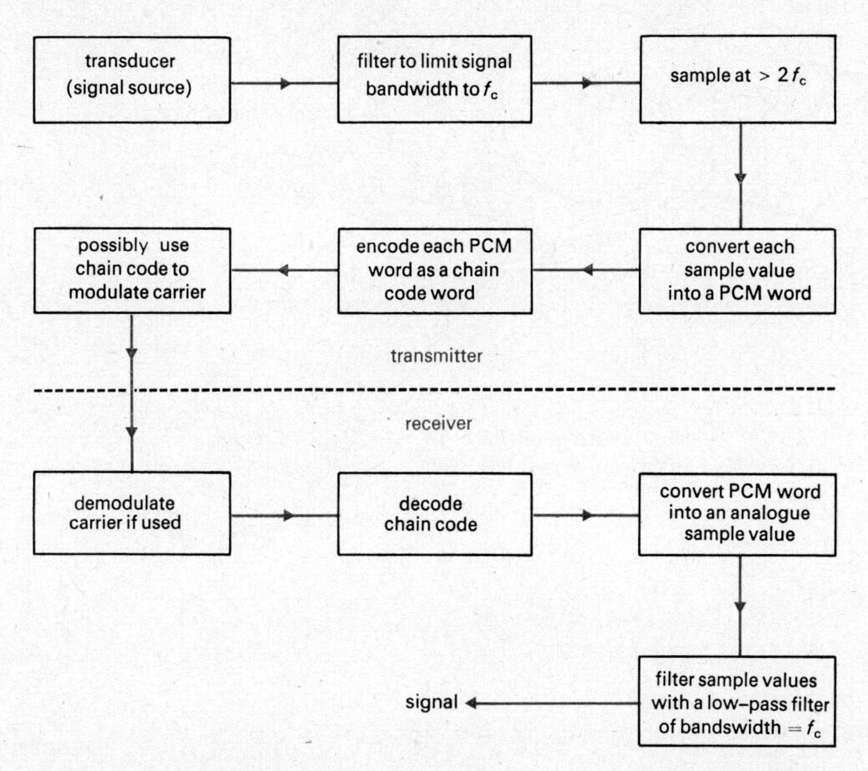

Figure 9 Schematic diagram of an instrumentation system which uses chain codes

SAQ 10

SAQ 10

In this example of transmitting five-digit numbers by means of chain codes, by how much has the bandwidth been increased in order to achieve this error-correction coding?

In the next section we shall consider some practical realizations of these ideas.

The Mariner 9 space probe

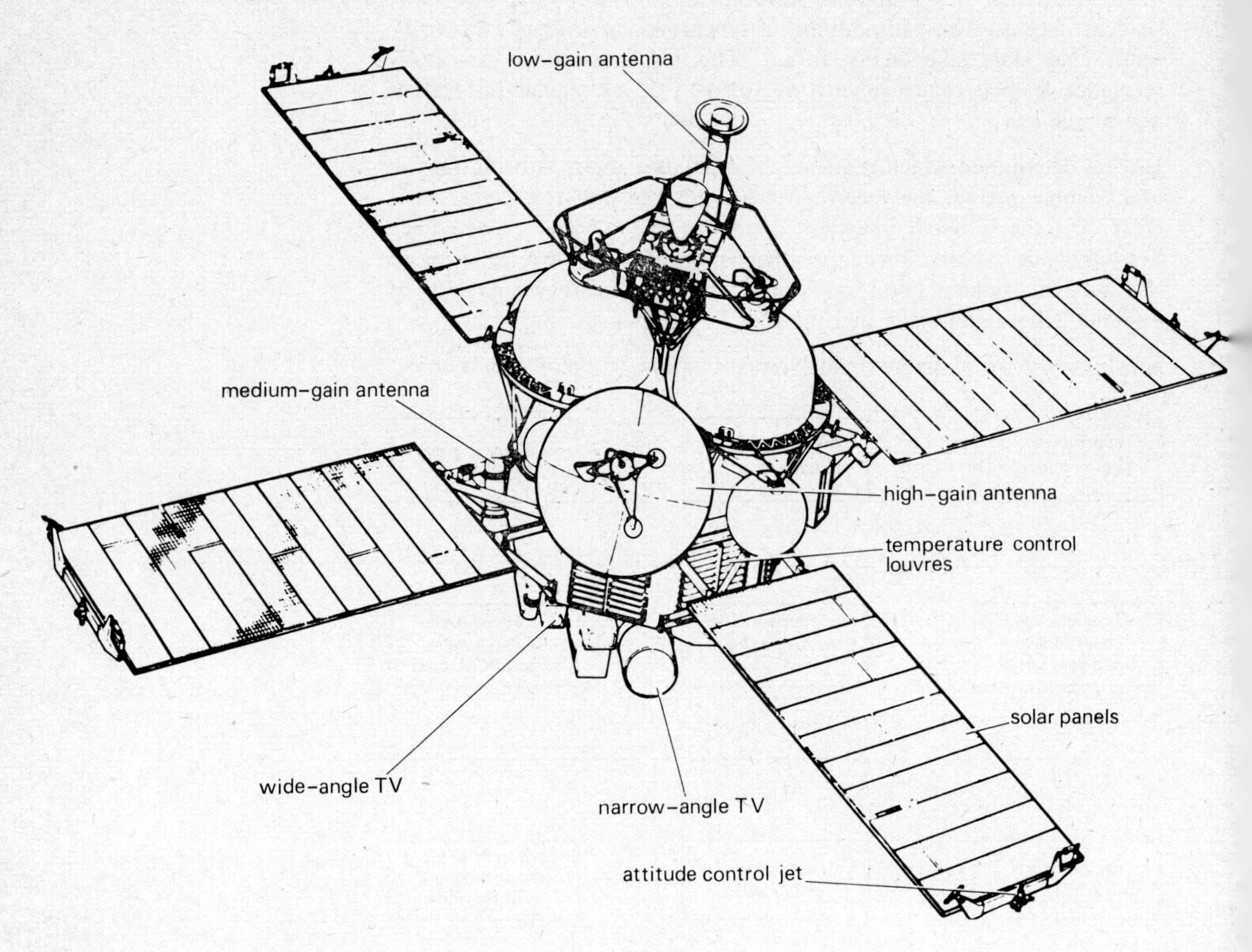

Figure 10 The Mariner 9 spacecraft

Mariner 9 was a member of the Mariner space probe programme and was one of the most successful in terms of the amount of data and pictures it sent back to Earth.

Earlier Mariner probes had flown close by the planets Venus (Mariners 2 and 5) and Mars (Mariners 4, 6 and 7) but Mariner 9 was actually placed into orbit around Mars on 13 November 1971. Its journey had taken almost six months and during that time the probe had travelled 4.62×10^{11} m (287 million miles). In orbit, the probe sent back scientific data and television pictures of the planet's surface for nearly a year. Then the spacecraft's transmitters were shut down upon a command from Earth because the fuel supplies for the jets which controlled the attitude of the space probe were exhausted.

During its operational period Mariner 9 sent back about seven thousand pictures of the surface of Mars and approximately twenty times the combined volume of scientific data that had been transmitted by all the previous probes to Mars. Two of Mariner 9's pictures are shown in Figure 11. Both pictures were taken from a distance of 1950 km. The picture in Figure 11(a) was taken with the wide-angle camera. The one in Figure 11(b) was taken with the narrow-angle camera and it shows the region marked by the rectangle in Figure 11(a).

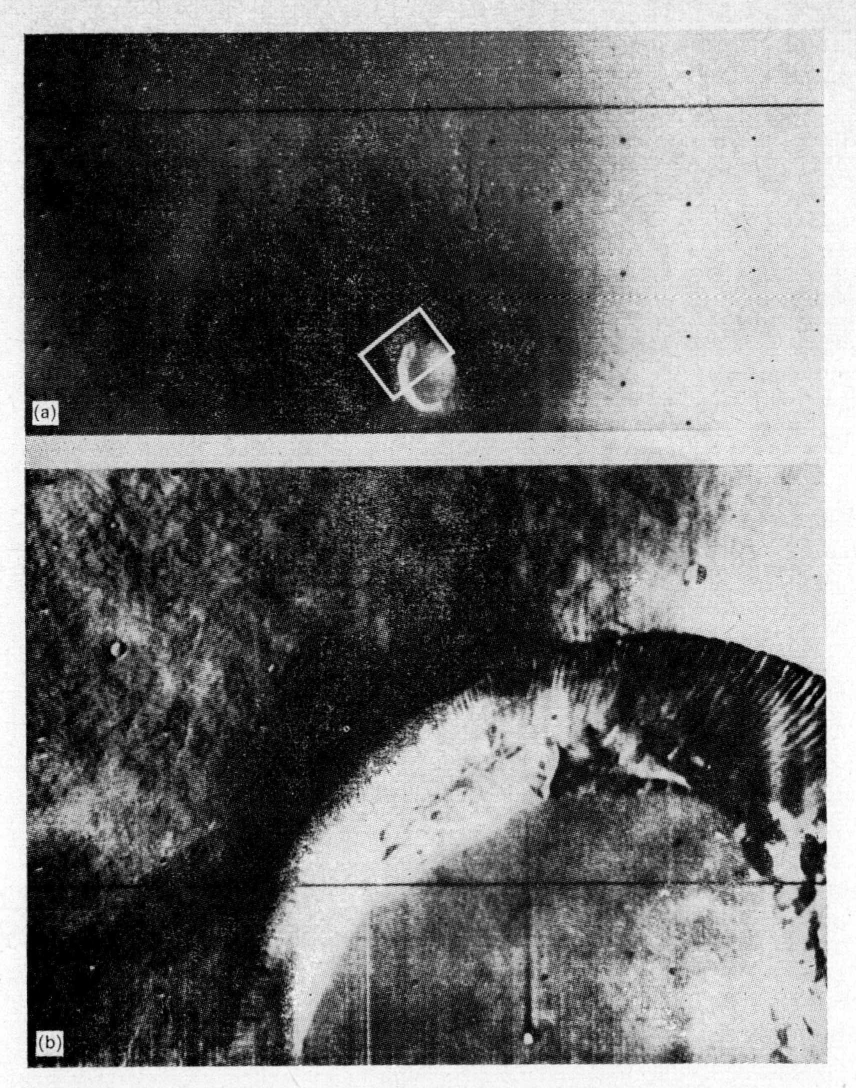

Figure 11 Two photographs of the surface of Mars taken by the Mariner 9 spacecraft from a distance of 1950 km*: (a) was taken with the wide-angle camera; (b) was taken by the narrow-angle camera. The width of the crater is* 45 km

As you can imagine, the probe was packed full of scientific instruments and transducers, many of which were similar to ones described in earlier units, for example, thermistors to measure the temperature of vital parts of the spacecraft, such as the temperature of the radio transmitters' amplifiers.

In this section I want to concentrate on describing two of the ways in which the Mariner system (space probe plus ground-support system) used error-correcting codes and pseudo-random binary sequences. The error-correcting code was used in connection with the communication system used by the space probe to return scientific data to earth. Pseudo-random binary sequences were used to make an accurate measurement of the distance of the space probe from the Earth. Without a knowledge of this distance the data and television pictures could not have been interpreted properly.

5.1 Mariner 9's use of error-correcting codes

In addition to describing the type of coding used by Mariner 9, I want first to outline some of the reasons why Mariner 9 used error-correcting codes.

Mariner 9 was about 1.4×10^{11} m (84 million miles) from Earth when it went into orbit. Its available transmitter power was 20 W. If all this power was radiated from an aerial which had no directional properties, the power would radiate out in all directions. An aerial on Earth of area 10 m^2 would receive only a tiny fraction of the transmitted power. This fraction is equal to the area of the receiver's aerial expressed as a fraction of the area of a sphere of radius 8.4×10^{11} m. This fraction is

$$\frac{10}{4\pi \times (8.4 \times 10^{11})^2} \approx 1 \times 10^{-24}.$$

Thus the received power would be of the order of $20 \times 1 \times 10^{-24}$ W.

A directional aerial on the spacecraft or a larger aerial on Earth could increase this by a few orders of magnitude, but still the received signal power would be very small indeed. The larger this signal is in comparison to the interfering interstellar noise, the easier it is to distinguish it from the noise.

The method used, in earlier Mariner probes, to ensure that the signal could be easily recovered in the presence of noise was to transmit the data in digital form, and to transmit the digits slowly.

Why does this improve things?

Unit 12 section 5.3 demonstrates that the amount of bandwidth required to transmit a PCM signal is directly proportional to the speed at which the digits are sent. This being the case, the bandwidth of the signal will be less for digits which are sent slowly than for digits transmitted at a faster rate.

The main source of noise interfering with the transmissions of the Mariner probe was interstellar noise, which has a uniform power density spectrum over the range of frequencies used by the probe's transmitters. Thus, if the channel bandwidth is decreased, the bandwidth of the interfering noise decreases and so does the total power in the interfering noise.

A data rate of 8.3 bits per second was used in these earlier probes. They flew past the planet under investigation and recorded the data on tape. The recorder was then played back slowly in order to transmit the data to the ground station. The tape recorder on the space probe was actually replayed twice. This is an example of the first type of error-detecting code mentioned in this unit.

However, Mariner 9 went into orbit around the planet and its instruments were capable of providing important data continuously. Also, the rate at which its instruments produced data was higher, so much so that the probe's tape recorder had to be able to record at 132 000 bits per second just to store the coded samples of the pictures being taken.

To cope with these increased data, Mariner 9 was designed to be able to transmit 16 200 message bits per second.

How much more bandwidth would this faster data rate require as compared with the 8.3 message bits per second?

Approximately 2000 times as much, that is, $16.2 \times 10^3 \div 8.3$.

This means that the interfering noise power would be approximately 2000 times larger, because interstellar noise has a uniform power density spectrum up to very high frequencies. Mariner 9 did have a larger transmitter power (20 W instead of 10 W) and better aerials available, but even so the probability of the interfering noise causing errors in the received digits sent at this rate would have been so large as to make the data meaningless.

The error rate at the transmission rate of 8.3 bits per second used by the previous probes was about 1 in 10^4. The error rate that could have been expected if 16 000 bits were transmitted each second can be calculated using the technique described in the radio programme 'Signal statistics'. The calculation gives the value for this expected error as about 1 in 5. As each message word from Mariner 9 was six binary digits long, there would have been, on average, an error in almost every one of the received message words. So a technique capable of reducing the error rate was essential.

The type of code that was chosen was similar to the ones described earlier in the unit. The coding was implemented by taking the six-digit PCM message word, representing, for example, a sample value of an instrument reading or a sample of a television picture and converting it into what is called *(32,6) biorthogonal code.*

(32,6) biorthogonal code

'Biorthogonal' describes the values of cross-correlation coefficients that can be obtained between the code words. A (32,6) code is one in which a message word six digits long is converted into a code word thirty-two digits long.

How long would a chain code corresponding to a six-bit message code have been?

It would be sixty-three digits long (look back at section 3.2.2).

This (32,6) code is obviously somewhat different. So let me first describe how each code word in a (32,6) biorthogonal code can be formed.

The first data symbol of the message code word plays an important role in the encoding procedure. Not only does it become the first digit in the code word but it controls a switch within the encoder which determines whether a code or its complement is sent. The complement of a code is one in which all 1s in the code are changed to 0s and all 0s in the code are changed to 1s.

Suppose the encoder has to encode the message code word **0**01110 (the most significant digit is 0) then the five least significant digits are placed in a shift register of a chain-code generator as shown in Figure 12(a). The shift register is then clocked around and the output is a chain code word, as was described previously, with thirty-one digits in it.

The resulting code word for 001110 is 0 followed by the sequence of thirty-one digits from the output of the shift register. This gives a total of thirty-two digits in the code word.

How many 1s and 0s are there in each code sequence of thirty-two digits?

There must be sixteen of each, because the chain code itself has fifteen 0s and sixteen 1s.

Now suppose the message code word is **1**01110. The 1 becomes the first digit in the new code word and also controls the logic switch in the

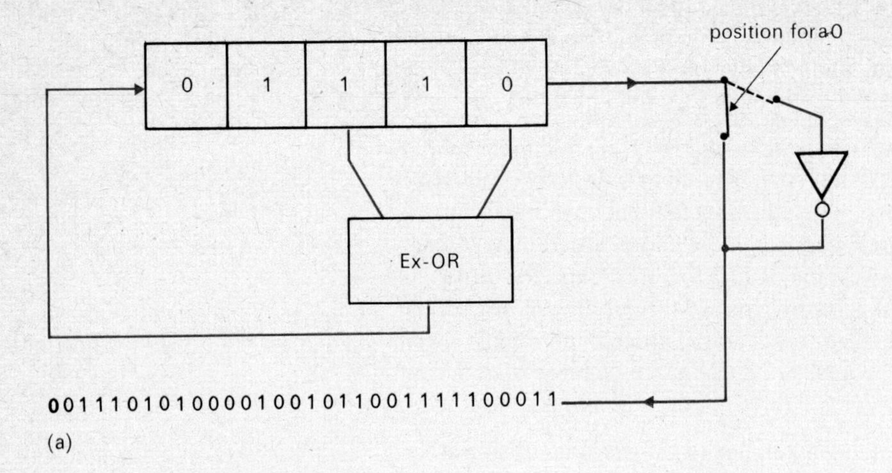

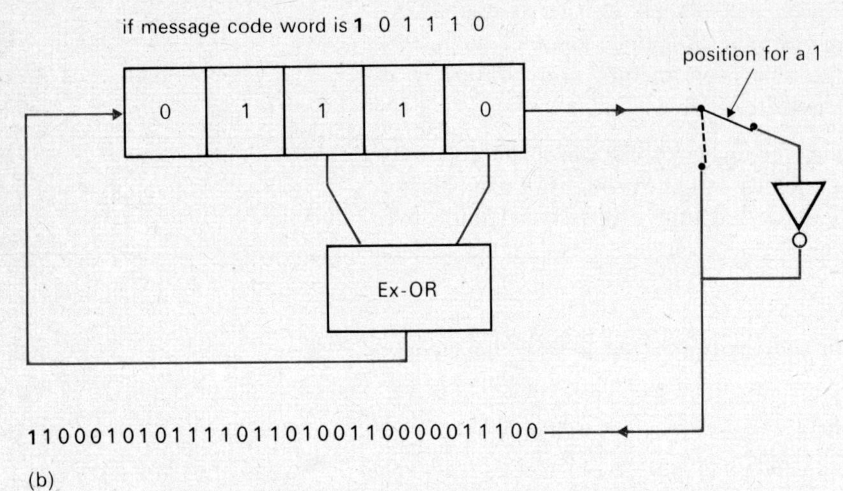

Figure 12 One method of producing the code words for a (32,6) biorthogonal code: (a) the most significant digit is a 0; *(b) it is a* 1

encoder so that the output of the shift register passes through an inverter before leaving the encoder. The code word for this message code is shown in Figure 12(b).

Notice that this code word is the exact complement of the code word for 001110. This means that if these two words were cross-correlated, their cross-correlation coefficient would be -1. I shall say more about this in a moment. Also notice that both kinds of code have equal numbers of 1s and 0s. This has practical advantages over the chain code, in which 1s exceed 0s.

In the Mariner space probe there were sixty-four different message words corresponding to the sixty-four (2^6) ways of arranging six binary message digits. Each one of these message code words could be converted into a (32,6) biorthogonal code word using a procedure similar to that outlined above, although the actual sequence of digits was somewhat different. Thus there were sixty-four redundant code words. These redundant code words were then modulated onto a carrier and transmitted to Earth.

On Earth the carrier was demodulated and the redundant code word, which of course had been affected by the interstellar noise, was obtained. The next step was to decode these code words, even though they contained errors.

The decoding procedure involved calculating cross-correlation coefficients between the received code words and stored versions of the code words. The stored code words need only be those generated in the situation shown in Figure 12(a). If this is the case, a correlation coefficient of $+1$ or -1 will be achieved between an error-free received code and one of the comparison codes. This is shown in Tables 9 and 10. All other cross-correlation coefficients will be zero, for example the one shown in Table 11.

So to decide which code word was sent, the decoder first finds which cross-correlation coefficient has the largest magnitude (i.e. which coefficient is largest if the $+$ or $-$ is neglected). Then, if the cross-correlation is positive, the decoder decides that the stored code which has given this large cross-correlation coefficient is the one that was transmitted. If the correlation is negative the decoder decides that the complement of the code word used for the comparison was the one that was transmitted.

If there are errors in the received signal, the effect of each one will be to reduce the magnitude of the correlation with the correct code word by $\frac{2}{32}$ ($1 \rightarrow \frac{30}{32}$ or $-1 \rightarrow -\frac{30}{32}$). The decoder working on the principle outlined in the previous paragraphs would be able to tolerate up to seven errors. This is because seven errors would cause the correlation with the 'correct' code word to be $\frac{18}{32}$ or $-\frac{18}{32}$, whilst with an incorrect code word it could be $\frac{14}{32}$ or $-\frac{14}{32}$. As the magnitude of the correlation with the 'correct' code word is greater than with any other, the decoder will function satisfactorily.

SAQ 11

SAQ 11

How does this (32,6) code compare with the chain code thirty-one digits long:

(a) in the number of code words;
(b) in error-correcting capability?

The error-correcting capacity of the Mariner code and the code word of thirty-one digits is the same. However, the Mariner code has about twice as many code words as the comparable chain code of thirty-one digits. This means that twice as many different message codes can be encoded using the Mariner code compared with the chain code of thirty-one digits.

Thus the Mariner code is an improvement on the straight chain code. This improvement is obtained by using both code words and their complements to encode the message code words.

The initial digit in the Mariner code seems to be unnecessary and, indeed, a similar improvement could have been obtained by using the code words of thirty-one digits and their complement. This initial digit does, however, ensure that there are equal numbers of 1s and 0s in each code word, as mentioned previously.

It is worth while to summarize the method used by the Mariner 9 space probe.

1 Data digits had to be transmitted at 16 200 bits per second.

2 These data digits were grouped in six-digit message words.

3 Each one of these message words was converted into a code word of thirty-two digits, the (32,6) biorthogonal code.

4 These code words were then transmitted to Earth. The rate at which these code word digits were transmitted was $16\,200 \times 32 \div 6 = 86\,400$ bits per second.

5 On Earth the receiver cross-correlated the received code words with a set of stored code words so as to determine which code was sent. This gives a capability of correcting seven errors for each of sixty-two possible message codes.

Table 9

Error-free received code	−1	−1	1	1	1	−1	1	−1	1	−1	−1	−1	−1	1	−1	−1	1	−1	1	1	−1	−1	1	1	1	1	1	−1	−1	−1	1	1
Stored code	−1	−1	1	1	1	−1	1	−1	1	−1	−1	−1	−1	1	−1	−1	1	−1	1	1	−1	−1	1	1	1	1	1	−1	−1	−1	1	1
Products	1	1	1	1	1	1	1	1	1	1	1	1	1	1	1	1	1	1	1	1	1	1	1	1	1	1	1	1	1	1	1	1

Average=1

Table 10

Error-free received code	1	1	−1	−1	−1	1	−1	1	−1	1	1	1	1	−1	1	1	−1	1	−1	−1	1	1	−1	−1	−1	−1	−1	1	1	1	−1	−1
Stored code	−1	−1	1	1	1	−1	1	−1	1	−1	−1	−1	−1	1	−1	−1	1	−1	1	1	−1	−1	1	1	1	1	1	−1	−1	−1	1	1
Products	−1	−1	−1	−1	−1	−1	−1	−1	−1	−1	−1	−1	−1	−1	−1	−1	−1	−1	−1	−1	−1	−1	−1	−1	−1	−1	−1	−1	−1	−1	−1	−1

Average=−1

Table 11

Error-free received code	−1	−1	1	1	1	−1	1	−1	1	−1	−1	−1	−1	1	−1	−1	1	−1	1	1	−1	−1	1	1	1	1	1	−1	−1	−1	1	1
Stored code	−1	1	1	−1	1	−1	1	−1	−1	−1	−1	1	−1	−1	1	−1	1	1	−1	−1	1	1	1	1	1	−1	−1	−1	1	1	−1	1
Products	1	−1	1	−1	1	1	1	1	−1	1	1	−1	1	−1	−1	1	1	−1	−1	−1	−1	−1	1	1	1	−1	−1	1	−1	−1	−1	1

Average=0

The overall effect of the Mariner error-correcting code was that the error rate at the output of the decoder, for the binary digits transmitted at the rate of 16 200 bits per second, was 1 in 200. (Compare this rate with the error rate of 1 in 5 that was predicted earlier if no error-correcting code was used.)

So the Mariner coding scheme is an example of the way in which a communication system had to be designed to complement the instrumentation system. In this case the instruments were producing a large amount of data during each orbit. To communicate the data to Earth the spacecraft had to transmit at a fast data rate. This led to too great a bit error rate but, by the use of an error-correcting code, many of these errors were corrected, producing an acceptable error rate in the decoded data.

5.2 The ranging of Mariner 9

The final use of pseudo-random sequences that I want to talk about was in determining the distance of the probe from the Earth. This distance is referred to as the *range* of the probe.

The method is basically very simple and extremely accurate. The principle of the method is that an extremely long pseudo-random sequence is transmitted from the ground station. This is received by the probe and retransmitted back to Earth. The round trip takes about seventeen minutes and during this time the pseudo-random sequence generator on the Earth will have moved on in its sequence.

The method of finding the delay produced by the round trip is illustrated in Figure 13.

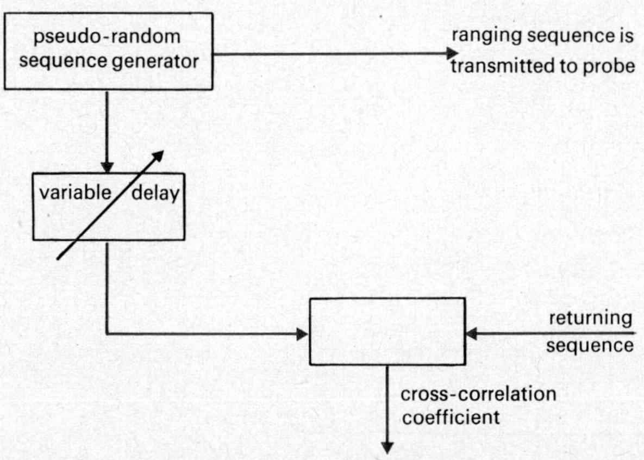

Figure 13 Schematic diagram showing the principle of the Mariner ranging system. The cross-correlation coefficient between segments of the two sequences is a maximum when the variable delay equals the round-trip delay

The output of the pseudo-random sequence generator on the Earth is passed through a delay circuit and then the cross-correlation coefficient between a segment of the returning sequence and a similar length segment of the sequence from the delay circuit is calculated. The value of the delay that has to be introduced on the delay circuit to get the largest cross-correlation coefficient is equal to the delay produced by the round trip to the space probe and back.

Let this delay be T, then the distance travelled by the signal is

$$(\text{Velocity of the signal}) \times T$$

and the range of the space probe is

$$\text{Range} = \frac{(\text{velocity of the signal}) \times T}{2}.$$

The velocity of the signal is known very accurately (it is the velocity of light) and with this method T can be determined accurately also. In fact, the range of the spacecraft was found to an accuracy of about ± 30 m at a distance from Earth to Mars of 1.35×10^{11} m (84 million miles). The *resolution* of the range readings was about 1.5 m in 10^{11} m.

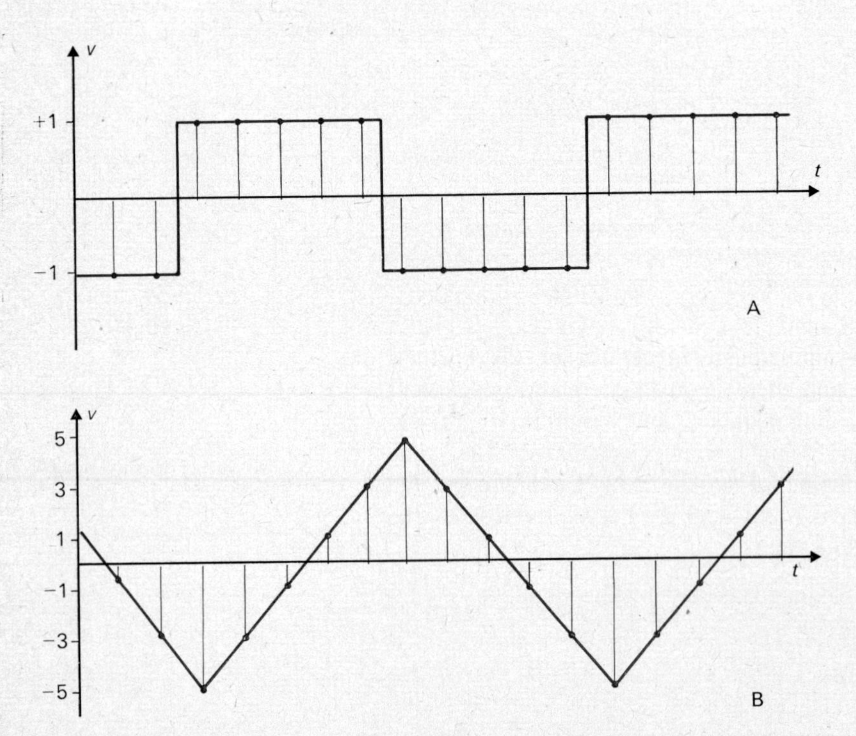

Figure 14 A sampled square wave and a sampled triangular wave

The cross-correlation function

This process of comparing one signal with another by multiplication of corresponding instantaneous values and taking the average is, as already indicated, called finding the cross-correlation coefficient. If next one signal is shifted in 'parametric time' with respect to the other and the coefficient found again, a graph of the value of the coefficient against parametric time can be plotted. This graph is called the cross-correlation function. It is a measure of the similarity between two signals, as we shall see, and has a variety of uses in measurement. In the following sections I want to show you (a) how it can be used to measure time delays and velocities (Section 7) and (b) how it can be used to measure the impulse response of a network or transducer (Section 8). But first let us be quite clear about what this function is.

The cross-correlation function differs from the autocorrelation function in that the autocorrelation function is a graph of correlation coefficients relating two versions of the *same* signal plotted against a shift in time (the so-called 'parametric time'), while the cross-correlation function is a graph of correlation coefficients relating two *different* signals plotted against the amount of time shift between them. Thus the autocorrelation function describes a particular signal, while the cross-correlation function describes the relationship between two signals.

We saw in Section 4 that we were able to use the calculation of the cross-correlation coefficients to decode chain codes which have errors in them. You may have noted a similarity in that process with the process for evaluating the autocorrelation function.

What property of chain codes gives rise to this similarity?

All the possible chain code sequences are the same sequence shifted in parametric time relative to each other. Thus, searching through all possible code words looking for the best fit is, in this case, the same as shifting one sequence step by step relative to the signal to be decoded.

Now let me explain in detail how the cross-correlation function is calculated.

We shall continue to consider sampled signals, rather than continuous signals, since it is easier to describe the process of calculating cross-correlation coefficients for sampled signals than for continuous ones.

Let us suppose that the two signals shown in Figure 14 are sampled at regular intervals. Within one period of the square wave, signal A, the sample values of the two signals A and B are shown in the first two rows of Table 12.

Table 12

A	+1	+1	+1	+1	+1	−1	−1	−1	−1	−1
B	−5	−3	−1	+1	+3	+5	+3	+1	−1	−3
Products	−5	−3	−1	+1	+3	−5	−3	−1	+1	+3

The pairs of simultaneous sample values have the products shown in the third row. The average of the products is exactly −1, so the cross-correlation *coefficient* for zero time-shift is −1. This is the first point in the graph of the cross-correlation *function*.

Now suppose waveform B is shifted one sample interval to the right. The sample values and products now become as shown in Table 13.

Table 13

A	+1	+1	+1	+1	+1	−1	−1	−1	−1	−1
B	−3	−5	−3	−1	+1	+3	+5	+3	+1	−1
Products	−3	−5	−3	−1	+1	−3	−5	−3	−1	+1

The average of the products this time is −2.2, so this is the cross-correlation coefficient for a parametric time shift of one interval τ_d.

SAQ 12

SAQ 12

Continue this process for different time shifts and plot the cross-correlation function of these two waveforms for values of between −5 and +5 intervals, that is, from $-5\tau_d$ to $+5\tau_d$.

In answering this self-assessment question you should have found that the function reaches its maximum for $\tau = -3\tau_d$.

What does this mean?

It means that in a quite particular sense these two different waveforms are 'more nearly the same' for $\tau = -3\tau_d$ than for any other value of τ. The function reaches its minimum value, within the range $+5\tau_d$ to $-5\tau_d$, at $\tau = +2\tau_d$ so, in the same sense, the two waveforms are most dissimilar for a time shift of $+2\tau_d$.

The value of the cross-correlation coefficient can be regarded as a *measure of similarity* between two waveforms. The cross-correlation function, among other things, enables us to find the parametric time shift which will achieve the maximum degree of similarity between two waveforms. The following case study on flow measurement describes one example of this.

Case study of flow measurement using correlation techniques

This section is concerned with a flow meter which uses cross-correlation techniques to determine the velocity flow rate of dirty liquids and slurries (mixtures of water and undissolved solids). The particular application we shall be examining is the measurement of the velocity flow rate and volume flow rate of sewage flows in a sewage-treatment works.

A modern sewage-treatment works is designed to treat domestic and industrial sewage so as to make it fit for discharge into a river or other watercourse. The sewage treatment is carried out in several stages, during which all of the larger particles are removed and the organic wastes in the sewage are degraded. The various processes involved in sewage treatment take place in a series of tanks, filters and activating beds. Raw sewage and partially treated sewage have to be pumped or allowed to flow between the various parts of the plant during the treatment processes.

During recent years a lot of interest has been shown in the automation of the sewage-treatment process. The reasons for this interest are that an automated treatment plant would produce a cleaner effluent and also automation of existing plants would enable them to deal effectively with larger quantities of raw sewage.

Adequate instrumentation is, of course, crucial to the automation of sewage treatment and one important instrument that will be needed is a flow meter capable of measuring the sewage flow rates.

Most of the flow meters mentioned in Units 8/9/10 can only be used with relatively clean fluids and they are therefore unsuitable for measuring flow rates of sewage, which can contain up to 10 per cent undissolved solids. Sewage with this percentage of undissolved solids is termed 'sludge'. Differential pressure flow meters, such as a Venturi flow meter, would get blocked too easily, by sludge being deposited at the orifice, and their use also involves a head loss.

What flow meter mentioned in Units 8/9/10 would be suitable for the measurement of sewage flows?

The electromagnetic flow meter would be suitable, since it causes little obstruction to the flow and therefore its use would not cause the pipe to block. In fact the electromagnetic flow meter has been used for flow measurements of sewage but it is relatively expensive and it has proved difficult to use.

Another type of flow meter has been developed, by Beck and others at the University of Bradford, which seems to be a promising way of measuring the flow rates of several fluids, including sewage. This flow meter uses cross-correlation techniques to determine the average time τ_f taken for the fluid to travel between two transducers a known distance D apart, as shown in Figure 15. Once τ_f is known the velocity flow rate of the fluid is given by D/τ_f.

The type of transducer which is particularly suitable for a cross-correlating sewage flow meter is a conductivity transducer. This indicates the

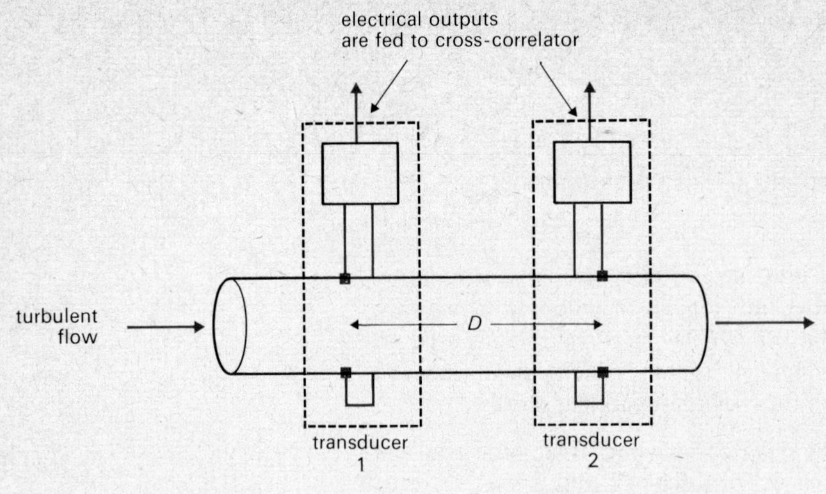

Figure 15 The position of the transducers in the cross-correlation flow meter

electrical conductivity of the sewage between two electrodes which are positioned across the pipe as shown in Figure 15.

The electrodes only need to protrude a small distance into the pipe, so there is little likelihood that they will cause the pipe to silt up and block. Commonly these electrodes are just the heads of bolts that have been bolted through the pipe as illustrated in Figure 16. They must, of course, be electrically insulated from the pipe if it is made from metal.

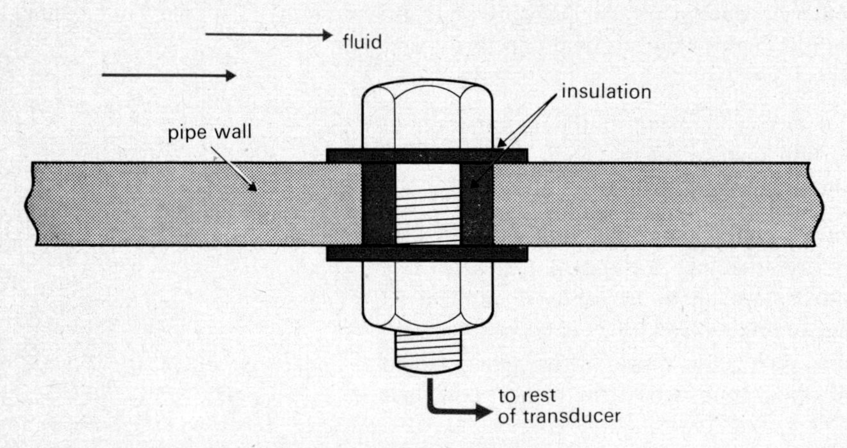

Figure 16 The bolt electrode of the conductivity transducer

The output of the transducers is an electrical signal which is proportional to the conductivity of the sewage between the electrodes. Now, the conductivity of the sewage varies with the concentration of suspended solids present: the greater the concentration, the larger is the conductivity. The concentration of suspended solids is not constant throughout the length of the flow since turbulence gives rise to pockets of less dense and more dense suspension, rather like clouds in the sky on a windy day. So the transducer output signals vary randomly because of the random concentrations produced by the turbulence. Two typical segments of the output signals are shown in Figures 17(a) and (b).

The two waveforms are not identical and yet we might expect them to be related to each other. The fluid that flows past the first set of electrodes goes on to flow past the second set. The distribution of solids will have altered somewhat while the fluid flows along, because of the turbulence present, but the output of the second transducer is related to a delayed version of the output from the first.

An instrument called a *cross-correlator* (sometimes shortened to *correlator*) is used to determine the delay time which gives maximum correspondence between the two waveforms.

cross-correlator
correlator

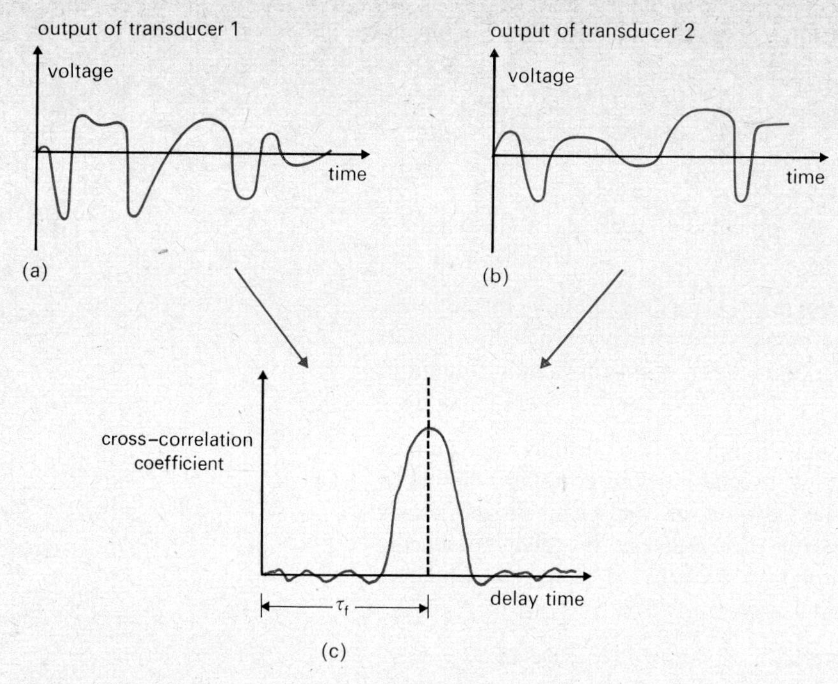

Figure 17(a) and (b) Typical waveforms from the conductivity transducers. (c) Shows the shape of the cross-correlation function of the waveforms in (a) and (b)

A cross-correlator is very similar to the autocorrelator described in Unit 12. A cross-correlator produces a plot of the cross-correlation function between two different waveforms.

For this flow meter the cross-correlation function that is obtained looks somewhat like that shown in Figure 17(c). The cross-correlator will, of course, only have a finite observation time, so it will only provide an estimate of the cross-correlation function of the signals. In a practical situation the observation time of the correlator is increased until a further increase produces negligible change in the cross-correlation function which is obtained. You can think of Figure 17(c) as being the result of this process.

The delay time corresponding to maximum correlation between the signals from the two transducers is the average time τ_f taken for the fluid to flow between the transducers. The velocity flow rate of the fluid is then D/τ_f, where D is the distance between the transducers. If the pipe is full and A is its cross-sectional area, the volume flow rate of the fluid is just $D \times A/\tau_f$.

So cross-correlation techniques permit the determination of the average transit time of sludge between two transducers a fixed distance apart, and hence the determination of the velocity of flow of the fluid.

7.1 Design problems

A number of questions arose in the construction of this flow meter and among the questions the designer had to deal with were:

1 Can the transducers be any distance apart?

In practice they can not, since the farther they are apart the less related signal 2 is to signal 1. This means that the cross-correlator has to examine longer portions of the signals in order to get a distinct peak in the cross-correlation function. Implied in this is that the measurements take longer and the cross-correlator is more expensive to make. Ultimately, if they are too far apart, no peak will be obtained.

The transducers cannot be too close together either since, if they are, there is a possibility of large percentage errors in the measurements of the distance D and delay time τ_f.

The usual compromise is to have the transducers spaced one pipe diameter apart.

2 How accurate is the flow meter?

Recent work on the accuracy has indicated that errors as low as ± 2 per cent can be obtained.

3 Most flow meters give the operator a reading of flow rate directly on an analogue or digital meter whereas, with the correlation flow meter, the operator has to interpret a plot of a cross-correlation function. Can this be overcome?

This is just a matter of choosing an appropriate display – a subject discussed more fully in Unit 15. A special-purpose correlator has been developed which can detect the *position* of the peak of the cross-correlation function. The correlator then converts the delay time corresponding to this peak position into a value of flow rate which is displayed on a meter in the normal way.

A simplified diagram of this correlator is shown in Figure 18.

The two signals to be correlated are fed into the comparator circuits 1 and 2. These comparators convert the analogue input signals into binary

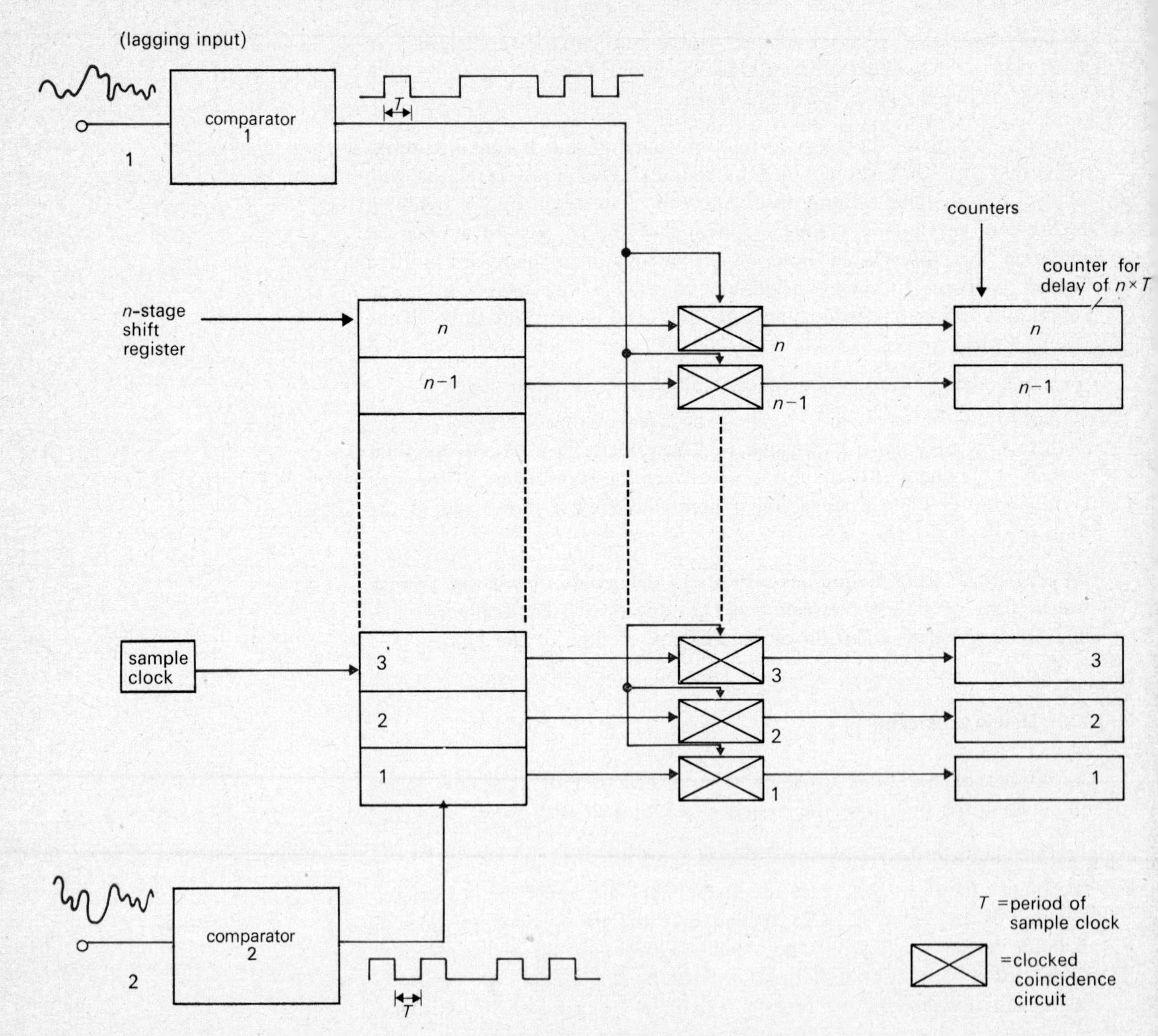

Figure 18 Simplified diagram of the special-purpose correlator devised for the cross-correlation flow meter

signals. The signals are sampled at intervals of T and whenever the sample is positive the output of the comparator is made equal to a voltage level corresponding to a 1, if the sample is negative, the output is made equal to a voltage level corresponding to a 0.

Comparator 1's output is fed to one input of each of the clocked coincidence circuits. These clocked coincidence circuits give an output of 1 if both of their inputs have the same value (i.e. both 1 or both 0) and a 0 if they are different each time a clock pulse occurs.

The other input to the correlator is also digitized and then fed digit by digit into an n-stage shift register which is shifted at intervals of T by the sample clock pulse whose period is T. The digits stored in the shift register are delayed versions of the signal fed into comparator 2. For instance, the value of the tenth stage will correspond to the value that was at the output of the comparator at a time $10T$ earlier.

So the basic components for a correlator are present: the shift register delays one signal and the clocked coincidence circuit makes the comparison. There is one of these comparison circuits for each delay that can be introduced by the shift register.

Thus the inputs to each of the coincidence circuits are the present value of signal 1 and one of the delayed values of signal 2. As mentioned previously, if the two inputs to a coincidence circuit are the same, the output from the circuit is a 1, when a clock pulse occurs. This 1 indicates a correspondence between the waveform 1 and a delayed version of waveform 2.

Now, the most 1s will be output from the coincidence circuit which is making a comparison between signal 1 and signal 2 delayed by a time which gives maximum correlation between the signals. So the number of 1s coming from the coincidence circuits are counted in individual counters, which are shown in Figure 18. The first counter to overload (i.e. to reach its maximum number and then return to 0) indicates the delay that gives maximum correlation between signals 1 and 2.

For example, if the first counter to overload is number 10, the delay time for maximum correlation is $10T$ and the velocity flow rate is given, as before, by $D/(10T)$. The resolution in this measurement is determined by the sample time T.

The correlator converts the information as to which counter overloaded first into a signal whose frequency is proportional to the reciprocal of the time delay for maximum correlation. A frequency-to-analogue converter and an analogue meter enables flow rates to be shown on a meter.

To summarize, the correlation flow meter uses the *variations* in transducer outputs at two places along the flow of a fluid. These output signals are cross-correlated and the peak in the cross-correlation function occurs at a delay time which equals the average time taken for the fluid to flow between the transducers. The distance between the transducers can be measured and the velocity of flow calculated.

SAQ 13

SAQ 13

The correlation flow meter described cannot be used to measure the flow rate of clean bubble-free water. Explain why this is so and try to think of some modifications which may make the measurement possible.

(*Hint*. Other types of transducer may be used also; remember that the flow meter depends on *variations* of some measurable parameter.)

Finally, this correlation flow meter can provide an estimate of the amount of suspended solids present in the fluid. This is because the size of the variations in output signals from the transducers depends on the concentration of suspended solids. The greater this concentration is, the greater the variations in the outputs from the transducers will be.

Another use of correlation is found in the measurement of the impulse response of a system.

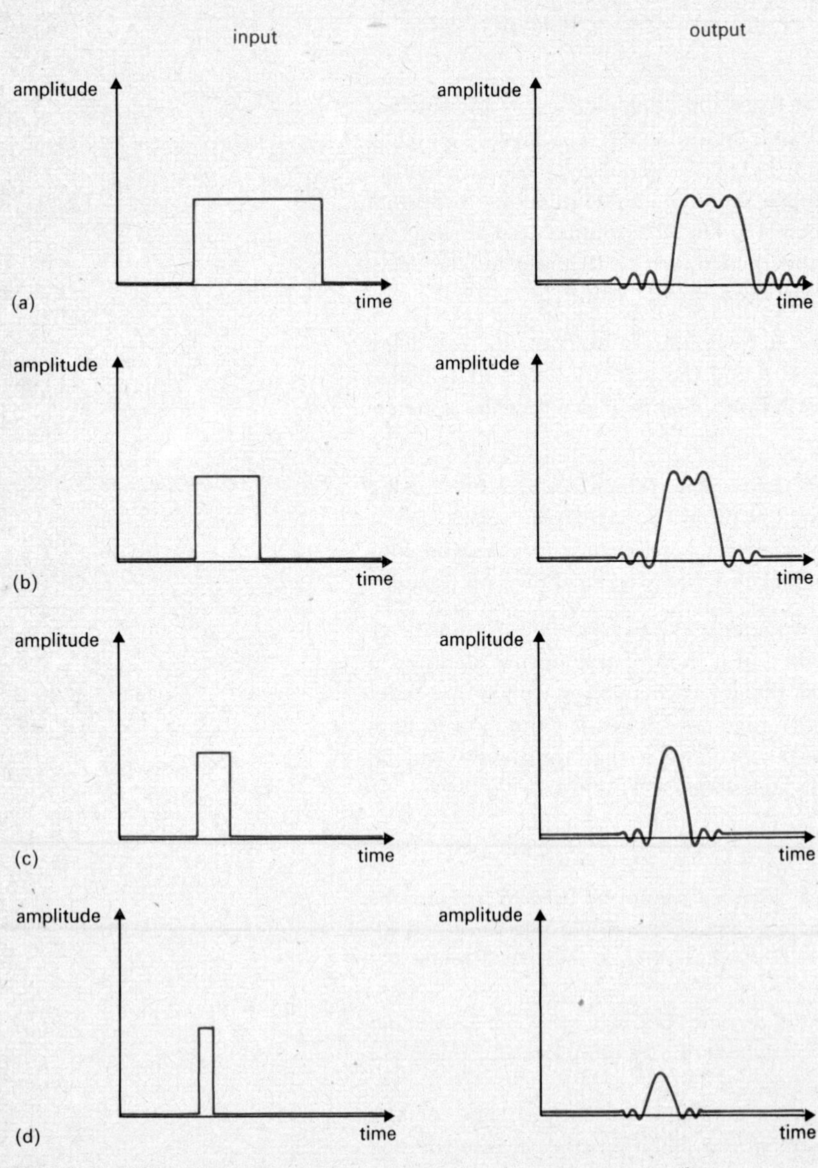

Figure 19 The response of a system to narrower and narrower input pulses

The measurement of the impulse response of a system using cross-correlation

The impulse response of a system is similar in many ways to the step response of a system discussed in Units 8/9/10. Both the impulse response and the step response completely characterize the system.

SAQ 14 (revision)

SAQ 14

Which other system measurement that you have learnt about also characterizes the system completely? How can this measurement be made?

The impulse response of a system can be measured by applying a narrow rectangular pulse to the input and observing the time variations of the signal at the output. Provided the pulse is narrow enough, the output will be the impulse response of the system to the required accuracy. An important question is: How narrow must this pulse be?

To answer this question let us consider Figure 19. The figure shows the output voltages of an electrical system which are caused by rectangular input pulses of different durations.

Notice that the size of the output decreases as the pulse width is reduced. Less energy is present in the input signal as the pulse width is reduced, so the output signal energy is correspondingly less. Also note that the shape of the output signal is the same for the two shortest input pulses. This is the impulse response of the system to the accuracy of the drawing.

The answer to the question of how narrow the pulse must be is that it must be so narrow that a further reduction produces a negligible change in the shape of the output from the system. This is because, as the pulse gets narrower, its spectrum broadens until, when it is narrow enough, it has a spectrum which is uniform, within the required accuracy, over the bandwidth of the system.

Impulse testing is a way of driving the system at all frequencies within its band simultaneously. The frequency response of the system can be found, from the impulse response, by a mathematical operation called the Fourier transform.

Impulse testing is also used to find the frequency response of systems with such a narrow bandwidth that a direct measurement of their frequency response would take a much longer time than an impulse test. Indeed, because of the longer time required to find the frequency response directly, the direct method may not be feasible in some situations. The method I have described for measuring the impulse response has the following limitation. If the system is 'noisy', that is, if there is an output signal even when no pulse has been applied to the input, then the impulse response will be degraded by this noise. Sometimes it is not possible to increase the amplitude of the pulse to overcome the effects of the noise, because a pulse of larger amplitude will cause the system to enter a non-linear region or limit.

Now, if it were possible to apply a signal with the same broad-band spectrum and containing the same energy but spread out in time, it ought to be possible to derive the impulse response without the same danger of overloading the network or driving it into non-linearity.

Random white noise is just such a signal and it can indeed be used to find the impulse response of a system.

How can this noise signal be used to determine the impulse response of a system?

The answer, which I shall state without attempting to prove, is that *the impulse response is given by the cross-correlation function of input and output waveforms.* This method is illustrated in Figure 20.

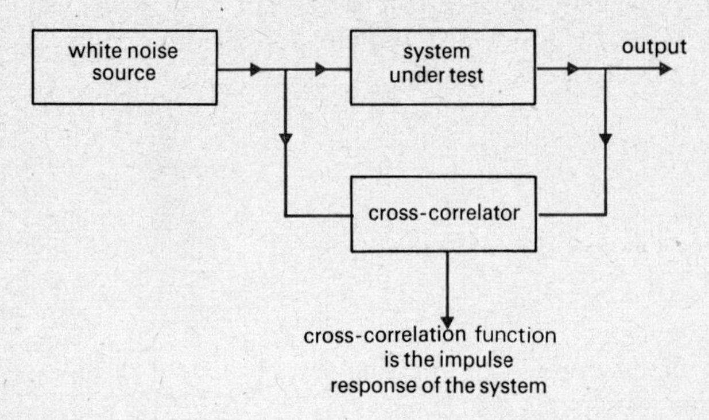

Figure 20 The measurement of the impulse response using cross-correlation

One of the important advantages of this method is that it can be used even when the system is in use for other purposes. The presence of other signals which have no correlation with the white noise produces no output from the cross-correlator and so does not affect the measured impulse response.

A pseudo-random binary sequence can be used as the input to the system, instead of natural white noise, provided it is chosen to have a power density spectrum which is uniform over the bandwidth of the system.

In fact, pseudo-random binary sequences are more often used than white noise from a 'natural' source.

One of the reasons for the use of pseudo-random binary sequences is that it is easier to perform cross-correlation if one of the signals to be correlated can only have one of two values. In addition the mean and peak power of a pseudo-random sequence can be easily found and this facilitates the choice of a sequence which does not overload the system.

8.1 Conclusion

We began this group of units by studying the properties of noise for the primary purpose of understanding how we can minimize its effect upon signals that interest us. Yet we have finished a study of its properties by finding that these very properties enable us to prefer it as a signal for certain kinds of measurement.

Self-assessment answers and comments

SAQ 1

(a) 4.

Each filter in spectrum analyser Y will pass four times the proportion of the spectrum passed by X. This means that the power in the waveforms from Y's filters will be four times that from X's.

(b) 1.

Power density readings allow for the differences in filter bandwidths.

SAQ 2

$-\frac{1}{3}$.

Table 14

Waveform	+1	+1	+1	−1	−1	−1	+1	+1	+1	−1	−1	−1
Two shifts	−1	−1	+1	+1	+1	−1	−1	−1	+1	+1	+1	−1
Products	−1	−1	+1	−1	−1	+1	−1	−1	+1	−1	−1	+1

Average of products is $-\frac{4}{12}=-\frac{1}{3}$.

SAQ 3

One possible circuit is shown in Figure 21. Check to see if it will work by finding its output for various values of the five inputs.

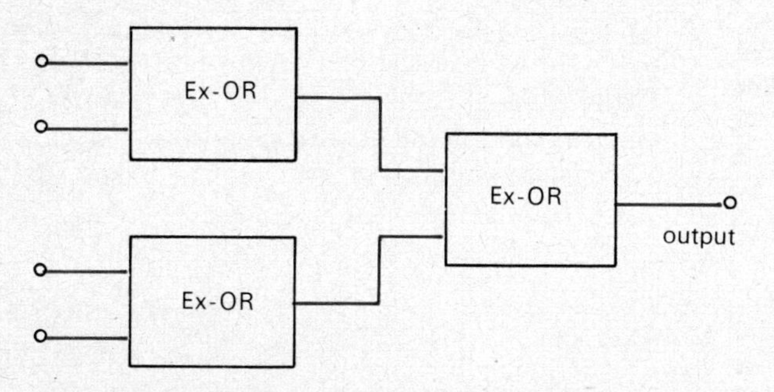

Figure 21 Answer to SAQ 3

SAQ 4

Remember that one of the resistor distributions considered in Unit 11 consisted of equal numbers of 11 Ω and 9 Ω resistors.

The Pascal's triangle that gave the relative frequency of occurrence of resistance values when the resistors were added together is repeated in Table 15.

Table 15

			1	1				10 Ω
		1	2	1				20 Ω
		1	3	3	1			30 Ω
	1	4	6	4	1			40 Ω
	1	5	10	10	5	1		50 Ω
1	6	15	20	15	6	1		60 Ω
1	7	21	35	35	21	7	1	70 Ω
77	75	73	71	69	67	65	63	

If you are not sure how this triangle is obtained, look back to Unit 11.

The numbers in red are the resistance values of the last row (nominally 70 Ω) while the numbers directly above are the relative frequency at which these values will occur and are equal to the number of ways a nominally 70 Ω resistor can be made by adding together seven resistors from the distribution.

Suppose that an 11 Ω resistor is equivalent to our binary 1 and that a 9 Ω resistor is equivalent to our binary 0. Then four ones and three zeros is equivalent to four 11 Ω and three 9 Ω resistors, that is, a total of 71 Ω. By looking at Pascal's triangle you should be able to see that there are thirty-five ways of obtaining a 71 Ω resistor from seven of these resistors. Similarly there must be thirty-five ways of obtaining four 1s and three 0s from seven digits.

SAQ 5

The answer is C.

Two errors can cause four different changes to the code. These are:

1 Two 1s are changed to two 0s.
First error, 1→0;
Second error, 1→0.
Resulting number of 1s=2.
Resulting number of 0s=5.

2 Two 0s are changed to two 1s.
First error, 0→1;
Second error, 0→1.
Resulting number of 1s=6.
Resulting number of 0s=1.

3 One 1 is changed to 0 and one 0 to 1.
First error, 1→0;
Second error, 0→1.
Resulting number of 1s=4.
Resulting number of 0s=3.

4 The same effect as in 3 is obtained if:
First error, 0→1;
Second error, 1→0.
Resulting number of 1s=4.
Resulting number of 0s=3.
The code can detect errors of types 1 and 2, since the resulting code contains a non-allowable number of 1s and 0s. The code cannot detect double errors of types 3 and 4, since these result in four 1s and three 0s.
Thus, the code will detect half of the possible error combinations.

SAQ 6

The chain code corresponding to a four-digit message 1000 is obtained by following the rules given for generating the four-digit chain code which were given in the text. This gives the answer

100010011010111.

This is the same sequence of digits as the code for 1101 but shifted along a few digits. Compare them to convince yourself of this fact.

SAQ 7

The only change in the rule for producing a chain code for a three-digit message is that you now perform modulo-2 addition on the first and third digit.

Starting with 111 modulo-2 addition of the first and third digit gives a 0 to be added to the end and the first digit is dropped. Modulo-2 addition is now performed on the new first and third digit. The whole procedure is repeated until the original signal is about to be regenerated.

The procedure is carried out below.

111
110
101
010
100
001
011
———
111

The code word consists of the first digit of all the three-digit patterns taken in sequence. It is equal to

1110100.

SAQ 8

The answer is seven.

A chain code of thirty-one digits contains sixteen 1s. Now, because modulo-2 addition between any pair of chain code words produces another chain code word containing sixteen 1s, there must be sixteen differences between any two chain code words of thirty-one digits.

If seven or less errors are present in the received code word, there will still be more differences between it and an incorrect code word than between it and the correct code word (nine as compared to seven).

For eight errors the number of differences between the received code word and the correct code word is eight. Between the received code word and an incorrect code word this number is also eight.

SAQ 9

The cross-correlation coefficient of a received code word with four errors with the correct code word would be $+1$ minus the effect of the errors causing a $+1$ (a correspondence) to be changed to a -1. Each error would reduce the correlation coefficient by $\frac{2}{15}$. The resulting coefficient would be

$$1-4\times\tfrac{2}{15}=\tfrac{7}{15}.$$

The maximum cross-correlation coefficient of the received code word with four errors with an incorrect code word would be $-\frac{1}{15}$ plus the effect of changing four -1s to $+1$s. Each error would increase the coefficient by $\frac{2}{15}$. The resulting coefficient would be:

$$-\tfrac{1}{15}+4\times\tfrac{2}{15}=\tfrac{7}{15}.$$

SAQ 10

Since the original measurand could be represented by samples taken at a rate of 100 per second, its bandwidth must have been less than 50 Hz (see Unit 11, section 5.1).

The final code involves a bit rate of 3100 bits per second. The 'highest-frequency' binary sequence is the pattern 1010 . . ., and this corresponds to a sinusoid of frequency $(3100\div2)$ Hz$=$ 1550 Hz.

Thus the bandwidth has increased by the factor $1550\div50$ just to transmit the error-correcting chain code.

The bandwidth must be increased a little more in order to provide for the synchronizing code sequence and still maintain the same data transmission rate.

SAQ 11

(a) A chain code thirty-one digits long has thirty-one chain code words as compared with the sixty-four code words in the Mariner code.

(b) Both codes can correct seven errors. Check back to section 4.2 if you are unsure about this.

SAQ 12

The cross-correlation function is plotted in Figure 22.

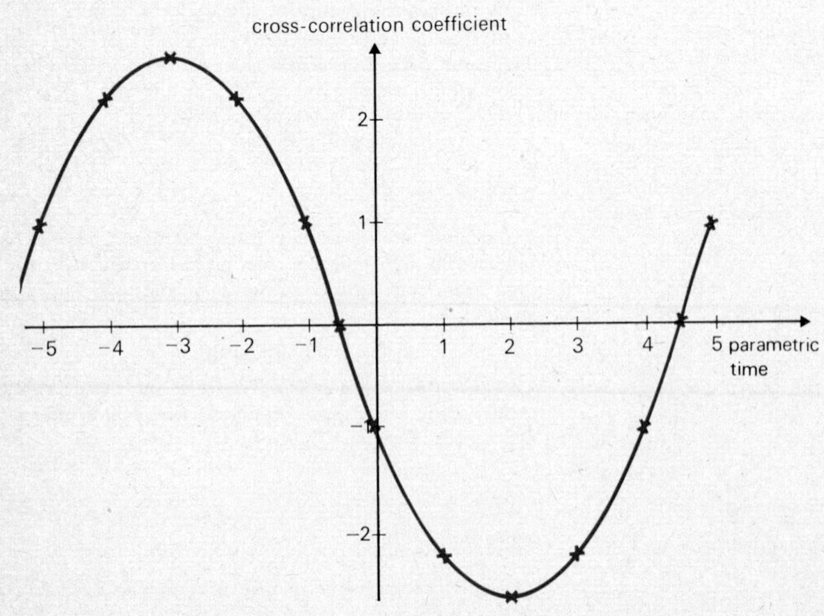

Figure 22 Answer to SAQ 12

SAQ 13

If the fluid were clean, the resistance between the transducer electrodes would remain constant. The transducers' outputs would be constant since they are designed only to output variations in conductivity. The output of the two transducers would then be a constant voltage as shown in Figure 23. Cross-correlation of these two signals would yield a flat cross-correlation function (see Figure 23c). Obviously this would be of no use in determining the average transit time of the fluid between the two transducers.

One possible modification to the apparatus would be to replace the conductivity transducer with temperature transducers and place a heater element up-stream of the transducers. If this heater element is controlled by a pseudo-random noise generator, it will introduce variations in the temperature of the fluid, these will be detected by the two transducers and the flow measurement can proceed as before.

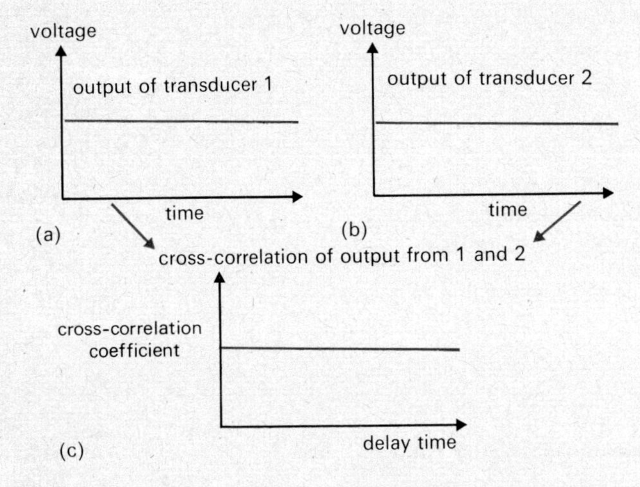

Figure 23 Answer to SAQ 13. (a) and (b) The outputs of the two conductivity transducers when the fluid in the pipe is clean. (c) The cross-correlation function of these two signals

SAQ 14

The frequency response of a system. You should remember from Units 8/9/10 that a system (an accelerometer) can be described by plotting the way in which it affects the amplitude and phase of a sinusoidal input for sinusoids of different frequencies. The measurement at each frequency needed to be performed separately and this meant that the measurement of the frequency response could be quite a lengthy procedure.

Summaries of Units 11, 12 and 13

Unit 11

This unit introduced the problem of receiving signals when noise or interference is added to them. The following principal topics were dealt with.

Quantifying a waveform

Any instrumentation system can deal only with a finite duration of a signal, the *observation time* T_o. Furthermore, it is always necessary to determine the *bandwidth* of a signal in order to quantify it. If this bandwidth is f_c, then $2f_cT_o$ quantities are required, in general, to specify the waveform. These quantities may be sample values, taken at a frequency of $2f_c$, or they may be Fourier components of the waveform. **2.2** **2.3** **5.1**

Random waveforms cannot be specified with fewer than $2f_cT_o$ quantities. **5.3** *Deterministic waveforms*, including periodic ones, may be specifiable with fewer than $2f_cT_o$ quantities. **5.3**

Average values

The mean, or average, value. If the sample values of a waveform are known, then the *mean* value of the waveform can be calculated by adding together all these values and dividing by the number of samples. **6.1**

If $x_1, x_2, x_3, \ldots, x_n$ are the values of n samples, then the mean value of x, written as x_m is given by

$$x_m = \frac{x_1 + x_2 + x_3 + \ldots + x_n}{n}.$$

The root mean square value. Using the same notation again, the r.m.s. value of x, written as x_{rms}, is **6.2**

$$x_{rms} = \sqrt{\frac{x_1^2 + x_2^2 + x_3^2 + \ldots + x_n^2}{n}}.$$

The standard deviation σ. This is a measure of the departure of the samples of a waveform from its mean value x_m. **6.4**

$$\sigma = \sqrt{\frac{(x_1 - x_m)^2 + (x_2 - x_m)^2 + \ldots + (x_n - x_m)^2}{n}}.$$

If the waveform has a *mean value of zero*, then its *standard deviation is equal to its r.m.s. value.*

The probability density function **6.5**

The probability density function for a segment of a signal is obtained by first plotting a histogram of the frequency of occurrence of sample values within each of a number of specified ranges of values against the midvalues of the ranges. Then the probability density function for this segment is obtained by drawing a smooth curve through the normalized histogram.

Alternatively, the probability density function may be derived theoretically from a knowledge of the origins of the waveform.

The Gaussian distribution

The Gaussian distribution always arises when *many distributions* of sample values are added together to form an overall probability density function. It is a bell-shaped distribution. Sixty-eight per cent of the distribution lies less than one standard deviation away from the mean (i.e. between $\pm\sigma$). About 95 per cent lies within $\pm 2\sigma$, and about 99 per cent lies within $\pm 3\sigma$. The latter two figures are the basis of the phrases '5 per cent confidence limit' and '1 per cent confidence limit'. **Section 7**

When n Gaussian distributions, each of standard deviation σ, are added together, the standard deviation of the sum is $\sigma\sqrt{n}$. If the *average standard deviation* σ_{average} is calculated, by dividing this sum by the number n of distributions added together, then the result is a *reduced* standard deviation.

7.1

That is,

$$\sigma_{\text{average}} = \frac{\sigma}{\sqrt{n}}.$$

Signal averaging

Section 8

This property of the reduction in average standard deviation of a random waveform, by repeated addition of segments of a signal plus noise, can be used *to separate the wanted signal from the noise.* The average of the recurring, wanted signal does not alter in this averaging process, while the standard deviation of the noise decreases. The signal thus 'emerges' from the noise.

The case study on the medical averager illustrates this last point.

Section 8

Unit 12

This unit continues with an explanation of the parameters which characterize a random signal.

The continuous spectrum

An *aperiodic* waveform of finite observation time can be fully specified by its Fourier components in terms of a *line spectrum* showing the amplitude and phase of each component. The *longer the portion* of the aperiodic signal that is observed, the *closer are the lines* in the spectrum.

Section 2

While the shape of the line spectrum of many deterministic waveforms tends to limit as the observation time is made larger, the shape of the line spectrum of a random waveform does not.

Power density spectrum

The frequency components of a random signal are characterized by the power density spectrum.

Section 2

The power density spectrum of a random signal can be found from the line spectra which describe segments of the signal or by using a spectrum analyser.

Signal filtering

Some examples were given of matching the system bandwidth to the bandwidth of the signal.

3.1

The autocorrelation function

The autocorrelation function is closely related to power density spectrum and, like power density, it contains no phase information about the signal.

Section 4

The process of autocorrelation establishes the relationship between the signal and a *time-shifted version* of itself. This relationship *for all possible time shifts* is described by the autocorrelation function.

Section 4

Uses:

1 By the examination of the autocorrelation function of a signal it is possible to see if it is aperiodic or periodic.

Section 4

2 The power density spectrum can be obtained from the autocorrelation function via the Wiener–Khintchine relation. In addition, information about the frequency content of a signal can be obtained from an inspection of the autocorrelation function without having to evaluate the Wiener–Khintchine integral.

Section 4

3 By examining the autocorrelation function of a composite signal it is sometimes possible to distinguish the separate parts of the autocorrelation function caused by the components of the composite signal. This enables in some cases the detection of a signal in the presence of noise. 4.1

Modulation

Towards the end of Unit 12 modulation is introduced as a means of modifying a signal before transmitting it over a channel so as to make it less vulnerable to the effects of noise. The three modulation schemes described were *amplitude modulation* (AM), *frequency modulation* (FM) and *pulse code modulation* (PCM). 5.1 5.2 5.3

In amplitude modulation the message signal modulates the amplitude of a carrier signal. In frequency modulation the message signal modulates the instantaneous frequency of a carrier. In pulse code modulation the message signal is sampled, and these sample values are converted into groups of binary pulses, each group representing one sample value as a binary number.

Frequency modulation and pulse code modulation enable the signal-to-noise ratio at the output of the demodulator to be increased by increasing the FM deviation or the number of digits in the PCM group. However, both frequency modulation and pulse code modulation require more bandwidth than amplitude modulation. 5.4

Unit 13

This unit continues with the theme of modifying or *processing* a signal before transmission so as to make the signal less susceptible to noise.

In particular, the unit describes techniques of *error detection* and *error correction* as applied to digital signals.

An *error-detecting code* is one which reveals to the receiver that an error has occurred during transmission. Section 2

An *error-correcting code* is a code which can automatically correct for a certain number of errors occurring in a code word. Section 3

The basis of all error-detecting and error-correcting techniques is the addition of *redundancy* to the signal. With a digital signal this means adding more digits to each binary signal in such a way that these extra digits are related to the original signal. Section 2

One example of an error-correcting code is the *chain code*. These are formed by using modulo-2 feedback on the appropriate stages of a shift register. 3.1

A set of chain code words can be represented by a circle or chain of digits. The feature which distinguishes one code word from another is its starting point in the chain. 3.2

The length of a chain code word is 2^n-1, where n is the number of digits in the message word. 3.2

The maximum number of errors that a chain code can correct is $2^{n-2}-1$. 4.2

The calculation of *cross-correlation coefficients* between a received code word and every possible code word is one way of decoding a chain code. The Mariner space probe gives an example of this technique. Section 4 5.1

The *cross-correlation function* between two waveforms is obtained by plotting values of cross-correlation coefficients of two waveforms against shifts in parametric time of one waveform with respect to the other. Section 6

The instrument which measures cross-correlation functions is called a *cross-correlator*. Such a cross-correlator is used in the measurement of *flow rates* and the measurement of the *impulse response* of a system. Section 7 Section 8

Acknowledgement

Unit 13

Grateful acknowledgement is made to NASA for Figures 11(a) and (b).

Instrumentation

1 Introduction to instrumentation
2 The measurement of strain
3, 4 Transducers 1: Temperature, displacement, force, torque, pressure
5, 6, 7 Numerical control of machine tools
8, 9, 10 Transducers 2: Acceleration, vibration, velocity, flow
11, 12, 13 Noise in instrumentation systems
14 Recording
15 Displays
16 Instrumentation in train development